Clare Feeney

HOW TO CHANGE THE WORLD

A practical guide to successful environmental training

SECOND EDITION

This edition published in 2019 by
Gosbrook Professional Publishing Ltd
20 Patrick Road
Reading RG4 8DD, UK

www.gosbrook.com

First published in 2013 by
Global Professional Publishing Ltd

Icons: tools ✂ by Maxim Kulikov from the Noun Project,
drawing ✍ by Daniel Shannon from the Noun Project

ISBN 978 1 912184 09 5 (paperback)
ISBN 978 1 912184 10 1 (hardback)
ISBN 978 1 912184 11 8 (ePub)
ISBN 978 1 912184 12 5 (PDF)

Cover and text design: Anke Ueberberg
Cover image: Matthias Hloucha

'... Clare makes environmental training accessible and impactful for you, the reader, and your audience. She successfully provides a framework for transitioning abstract ideas into practical realities. This book sets the standard as optical lens for environmental training programs.'

Dwane Jones PhD, Acting Dean of the College of Agriculture, Urban Sustainability, Environmental Sciences (CAUSES) and International Trainer for NGICP, University of the District of Columbia, Washington, DC

'At a time when evidence is coming thick and fast as to the environmental damage humans are inflicting on the planet to be able to sustain life, we need practical ways to respond. Adult learning and education, of which environmental education and training is a crucial part, are vital in our individual and collective responses. This manual supports trainers and educators to know what to do, while deepening their understandings of the root causes of the problem. I welcome warmly this very helpful resource and trust many will put it to urgent use!'

Professor Emerita Shirley Walters, Adult and Continuing Education, University of the Western Cape, South Africa

Praise for the first edition:

'Inspiring case-studies, lively writing and, best of all, a series of steps and tools to enable any organisation to improve its environmental performance. Clare's depth of experience and wisdom on good environmental practice and how to encourage people to commit to it shines through on every page.'

Niki Harré, Professor, School of Psychology, University of Auckland, and author of *Psychology for a better world: Working with people to save the planet*

Contents

Figures and tables

FIGURES

TABLES

Glossary

The following terms are defined for practical application. The definitions are not a substitute for academic or technical references.

avatar   Also known as a persona, an avatar is a fictitious character who represents one of the many different groups of real people who make up your target audiences. The purpose of developing an avatar is to make sure you understand the learning preferences and needs of your trainees.

(baby) boomers   This large demographic group has birth dates between the end of the Second World War and 1964 and has enjoyed relative affluence, compared with their parents' generation. Cf. **Generation X**.

baseline   A baseline is a clearly defined starting point from which implementation begins, improvement is judged or comparison is made.

benchmark   A benchmark is a level of performance generally accepted as best practice. Benchmarks may identify substandard, average and excellent levels of performance and are usually quantitative rather than qualitative, to allow for comparison across time. Standards can be used as benchmarks, especially where there is an element of compliance.

best practice   Best practice is the most effective and scientifically sound techniques known to deliver a desired outcome.

capability   Capability defines a ceiling on what a given group of people can achieve based on the knowledge, skills and experience of individuals or a **sector**. Capability also refers to potential – that is, those aptitudes that can be developed. Cf. **capacity**.

capacity   Capacity defines a ceiling on what a given group of people can achieve that is defined by their headcount, funding and access to other resources. Cf. **capability**.

course   A course comprises an integrated set of sequential and associated **training** activities on a given topic or set of topics that are delivered over a period of weeks or months, whatever the form of delivery, which may include face-to-face **workshops**, e-learning, site visits, field trips and more.

education   Education refers to the process of receiving or giving systematic instruction, especially at a school, technical college or university.

formative evaluation   A formative evaluation answers questions about how to improve and refine a developing programme. It is usually undertaken during the initial or establishment phase of a project, although it can also be helpful for assessing the ongoing activities of an established programme. Formative evaluation may include process and impact studies. Cf. **summative evaluation**.

generation X (Gen X)   The name given to the demographic cohort born between the early-to-mid 1960s and the early 1980s. Cf. **baby boomers**, **Generation Y**.

generation Y (Gen Y)   Also referred to as millennials, this generation has birth years between the early 1980s and the mid-1990s to early 2000s. They have come of age in the digital era and are comfortable with the internet, digital technologies and social media. Cf. **Generation X**, **Generation Z**.

generation Z (Gen Z)   This generation is typically considered to have birth years starting in the mid-1990s to early 2000s. They have used digital media from a young age. Cf. **Generation Y**.

green jobs   Green jobs are a vital part of the 'transformation of economies, enterprises, workplaces and labour markets into a sustainable, low-carbon economy providing decent work'.[1] The International Labour Organization (ILO) defines green jobs as decent jobs that: reduce consumption of energy and raw materials; limit greenhouse gas emissions; minimize waste and pollution; protect and restore ecosystems; and contribute to adapting to climate change.

industry   Industry is a term used to refer to a very specific group of companies or businesses within a **sector**.[2] For example, food manufacturers are part of the secondary sector and that sector may comprise many specialized industry associations. Such industry associations coordinate their members' activities and act as contact points.

indigenous capital   Every indigenous group will have their own definition of the knowledge and values that inform their sustainability aspirations and practices and how they wish them to be applied. For the purposes of this book I define indigenous capital as a seventh capital to and from which value can be added, in addition to the sometimes substantial stocks and flows of the financial and other capitals that indigenous groups may have. My definition is informed from the perspective of a pakeha (person of Caucasian descent) from Aotearoa New Zealand and is set out in more detail in Chapter 4.4.

influencers   Influencers are those people, groups or organizations with interests in a policy, programme or project whose support (or lack of it) might significantly impact on a project's success.

learning objectives   Also sometimes referred to as training objectives, these reflect the findings of a **training needs assessment (TNA)**, and objectively describe what knowledge the trainee will acquire and what specific practices the trainee will be able to carry out as a consequence of the training.

millennials   *See* **Generation X**.

needs assessment   *See* **training needs assessment (TNA)**.

---

1 Green Jobs Assessment Institutions Network (GAIN) (2017) *How to measure and model social and employment outcomes of climate and sustainable development policies: training guidebook.* International Labour Organization. Available at www.ilo.org/global/topics/green-jobs/areas-of-work/gain/training-guidebook/lang--en/index.htm [accessed 12 April 2019].

2 Chad Langager (2019) Industry vs. sector: what's the difference? Available at www.investopedia.com/ask/answers/05/industrysector.asp [accessed 19 September 2019].

pedagogy   Pedagogy refers to the principles and methods of enjoyable and effective instruction and the orderly and logical arrangement of learning content into a series of topic steps.

persona   *See* **avatar.**

productivity   The New Zealand Productivity Council defines productivity thus: it is 'how well people combine resources to produce goods and services. For countries, it is about creating more from available resources – such as raw materials, labour, skills, capital equipment, land, intellectual property, managerial capability and financial capital. With the right choices, higher production, higher value and higher incomes can be achieved for every hour worked', and we can reduce or avoid inputs of natural and other resources.[3]

sector   In this book, sector refers to broad groupings of economic activities. For example, the primary sector usually refers to extractive activities such as mining, forestry, fishing and agriculture; the secondary sector, to manufacturing of materials into goods; and the tertiary or service sector, to bodies that provide intangible services, including retail, consulting, financial and more. Two or three more sectors are sometimes also defined.[4] Business, trade or professional associations may coordinate the activities within a sector and act as contact points. Cf. **industry.**

social enterprise   A social enterprise is a commercial organization that has specific social objectives that serve its primary purpose. Social enterprises seek to maximize profits while maximizing benefits to society and the environment, and their profits are principally used to fund social programmes. They apply commercial strategies to maximize improvements in financial, social and environmental well-being. This may include maximizing social impact alongside making profits for external shareholders. Buying from social enterprises allows consumers to support positive social change and their community.

social licence   The phrase social licence is used to refer to the 'ongoing acceptance of a company or **industry**'s standard business practices and operating procedures by its employees, stakeholders and the general public. The concept of social licence is closely related to the concept of sustainability and the triple bottom line'.[5] This concept is now actively being measured by many **sectors** for sustainability purposes.

summative evaluation   A summative evaluation answers questions about programme quality and impact for the purposes of accountability and decision-making. It is conducted at a project's/programme's end and usually includes

---

3 New Zealand Productivity Commission (2013) Towards better local regulation. Available at www.productivity.govt.nz/inquiries/towards-better-local-regulation/ [accessed 25 October 2019].

4 Tejvan Pettinger (2017) Sectors of the economy. Available at www.economicshelp.org/blog/12436/concepts/sectors-economy/ [accessed 19 September 2019].

5 See the definition at www.investopedia.com/terms/s/social-license-slo.asp. See also Robert Quigley and James Baines (2014) *How to improve your social licence to operate: a New Zealand industry perspective.* Aquaculture Unit, Ministry for Primary Industries, Information Paper No. 2014/05. Available at www.mpi.govt.nz/dmsdocument/3995-how-to-improve-your-social-licence-to-operate-a-new-zealand-industry-perspective [accessed 11 August 2019].

a synthesis of process and impact or outcome evaluation components. Cf. **formative evaluation.**

sunrise industry Sunrise industry is a term used to refer to a **sector** or **industry** that is in its infancy, but which shows promise of a rapid boom, typically characterized by high growth rates, rapid innovation, numerous start-ups and the availability of venture capital funding. Renewable energy and other low-carbon initiatives, more sustainable foods and materials, waste avoidance and specialized green jobs are among emerging sunrise industries in the sustainability sector, as are social enterprises and other businesses focused on well-being. Cf. **sunset industry.**

sunset industry With respect to sustainability, a sunset industry can be defined as one whose effects on the natural environment and/or people mean that it needs to radically reduce its carbon intensity and other emissions into air, land and water, and/or to radically improve its ethical practices and human impacts to become more sustainable and equitable. Other definitions include: an **industry** in decline that has passed its peak or boom periods; an industry that has existed for a long time and which is less successful and making less profit than before; and an older industry that continues to be important to an economy, but is losing favour with investors because of its steadily falling employment generation capacity and profits, and comparatively higher environmental costs. Cf. **sunrise industry.**

training Training delivers work-related knowledge, skills and practices that will improve an accepted and specified aspect of on-the-job performance that can be defined and objectively assessed in observable and/or objectively measurable ways. Trainers, trainees and assessors all need to know exactly what is expected of them.

training needs assessment (TNA) A TNA is a way of determining if a **training** need exists and, if it does, what training is required to fill the gap. It is essentially an assessment of 'what is' and 'what ought to be' that defines the training and/or other needs. It is sometimes shortened to 'needs assessment' because training may not always be the solution to the problem.

training objectives *See* **learning objectives.**

training programme A training programme is specific to a given environment and sustainability topic. It defines **training** needs and the desired workplace performance, considers the trainees' learning needs and preferences, develops and delivers training to a planned time frame, and supports trainees and their up-line supervisors and managers to enhance application of new learning in the workplace. A training programme also sets out the indicators and methods for evaluating the outcomes of the training in terms of the trainees' desired performance, as well as the organizational and other outcomes set out in an overarching **training strategy.**

training strategy A training strategy sets out your approach to **training** in a way that is fully aligned with the wider external context within which your organization operates, its internal vision, mission, purpose and goals, and the risks and opportunities associated with both of these. It then aligns the purpose and outcomes of training with these based on best adult learning principles and practices in a way that can demonstrate the business value of investing in training. A training strategy may identify the need for several **training programmes** on different topics and will use clear criteria to prioritize these to show your managers that you're using your budget effectively.

workshop   A workshop is a method of delivering **training** that usually involves classroom-style interactive activities of a given duration (half a day, one day, two days, etc.). A workshop may be a standalone training method or it may be part of a **course** that additionally involves other delivery methods, including e-learning, field trips, webinars and the like.

Abbreviations

CEnvP	Certified Environmental Practitioner
CEO	chief executive officer
CPD	continuing professional development
CPESC	Certified Professional in Erosion and Sediment Control
CRM	customer relationship management
CRP	Certified ROI Professional®
DoJ	Department of Justice (US)
EPA	Environmental Protection Agency (US)
ESG	ecological, social and governance
GDP	gross domestic product
GI	green infrastructure
GRI	Global Reporting Initiative
IEMA	Institute of Environmental Management and Assessment
IIRC	International Integrated Reporting Council
ILO	International Labour Organization (UN)
IPR	intellectual property rights
ISO	International Organization for Standardization
ITUC	International Trade Union Confederation
KPIs	key performance indicators
L&D	learning and development
LMS	learning management system
MOOC	massive open online course
MOU	memorandum of understanding
NGICP	National Green Infrastructure Certification Program (US)
NPDES	National Pollutant Discharge Elimination System (US)
PR	public relations
ROI	return on investment
SDGs	Sustainable Development Goals (UN)
SME	subject matter expert
TNA	training needs assessment
TVET	technical and vocational education and training
UNEP	United Nations Environment Programme
UNESCO	United Nations Educational, Scientific and Cultural Organization
UNITAR	United Nations Institute for Training and Research
WBCSD	World Business Council for Sustainable Development
WEF	World Economic Forum

Preface

WHY THIS BOOK? WHY NOW?

Friends and colleagues asked me if this book was trying to do something bigger than environmental training. I'm so grateful: they helped me to articulate that it's not that the book is trying to be something bigger than it is, but that environmental training is so much bigger than people realize.

The powerful tools now available to help us to measure and monetize economic, social and environmental outcomes reveal that environmental training can deliver more change more easily than we thought. I've seen working people's lives enhanced by their new skills and companies radically increase their turnover as they embrace sustainability and organizational learning that delivers competitive advantage and wider social benefits.

Many years' experience and research into the transformational power of training have led me to wholeheartedly believe that environmental trainers really can help to save the world.

In writing the first edition of this book, I realized how very much the world of training and the world of environmental management have to offer each other.

My planetary training model fully integrates the worlds of professional training and environmental management: it's a breakthrough concept. The model aligns the measurable and monetized outcomes of environmental and sustainability training with the ways in which governments, stock exchanges, companies and not-for-profits measure the social, economic and environmental outcomes of their activities.

With this second edition of the book, I aim to elevate workforce training to its highest status: a powerful way of helping people to solve the serious environmental issues facing humanity, while at the same time solving their associated social and economic problems.

This is the most exciting thing I've ever done in my life!

WHO CAN BENEFIT?

People worldwide are creating entirely new professions of environmental and sustainability managers in every sector, adding tremendous value to businesses and communities. My work supports that of many international initiatives referenced throughout this book and its accompanying workbook. How will *your* sector benefit?

- **Businesses and utilities**   Use this book to help you to set up environmental training for your staff, contractors and suppliers to meet the increasingly detailed social and ecological standards that clients and customers are demanding. You can work with governments to design environmental training programmes to strengthen the relevance and effectiveness of the training and to enjoy the increased staff engagement and productivity that will result.

- **Government and regulatory agencies**   Use this book to work constructively with your stakeholders to set up environmental training programmes for firms in high-risk and high-compliance sectors as a cost-effective way of meeting your environmental management responsibilities. Global bodies, policy- and decision-makers are embracing environmental training to address the big issues of our time.

- **Professional, trades and industry associations, and workplace unions**   Take the initiative for the professional learning and development of your members with regard to the environment – and accrue the associated benefits in skills and engagement for workers and their employers. Be part of the 'just transition' of workers into green jobs and the creation of new green professions.

- **Environmental community groups, first (indigenous) peoples, environmental groups, not-for-profits and social enterprises**   You may have played a major role in environmental improvement in the past and may now see opportunities for growing your capability and capacity via environmental initiatives and their flow-on social and economic benefits.

- **Tertiary educators and professional trainers**   Your skills will be in demand from people wanting to work with this book.

This new demand for environmental and sustainability management professionals means new opportunities to make a very real difference in the world. Join us!

DISCLAIMER

The core principles of environmental training are generic to any environmental jurisdiction, but because legal and administrative frameworks and levels of programme resourcing vary from place to place, you will need to adapt the information provided to meet your own needs and opportunities.

I can't guarantee replicable results if you follow the advice in this book – but if you want to set up or enhance your environmental training programme or better evaluate its outcomes, it will certainly help you along your way.

ACKNOWLEDGEMENTS

I gratefully acknowledge permission from the staff of Water New Zealand to include in this second edition of the book material from their Stormwater Education, Training and Sector Development Plan – thank you for the opportunity to work with you. Many thanks also to the volunteer members of the Water New Zealand Stormwater Committee who supported this work, especially past and current committee chairs Mike Hannah of Stormwater 360 and James Reddish of WSP.

Likewise, the Healthy Waters Team at Auckland Council and Dukessa Blackburn-Huettner, the team's head of lifecycle management, kindly allowed me to include key learnings from Dukessa's move to introduce the National Green Infrastructure Certification Program (NGICP) from the US into New Zealand. It's a wonderful initiative.

Warm acknowledgement is also due to the people who inspired the first edition of this book: my co-trainer, Brian Handyside of Erosion Management Ltd, and the leaders of the Erosion and Sediment Control Programme run by the Auckland Regional Council and its successor, the Auckland Council, Kerry Connolly, Mark Cleaver, Earl Shaver, Graeme Ridley, Paul Metcalf, Mike Dunphy and Roger Bannister, as well as Stormwater Education and Community Programmes adviser Marcus Ballantyne. Brian and I learned a lot and had a wonderful time working with the equally gifted staff of many other councils in New Zealand, including Northland, Bay of Plenty, Waikato, Wellington, Canterbury and Dunedin.

A big thank you to Jay Wilson of the City of Charlotte, NC, for allowing me to cite his erosion and sediment control training programme as a detailed case study. More thanks to the people who allowed me to cite their work as other case studies and who gave very helpful feedback and advice, including Fiona Mountfort of Mountfort Consulting, Rich Batiuk of Coastwise Partners, Troy Brockbank, Kylie Eltham of Eltham Environmental, Amanda Davies, Michael Lindgreen of 4Sight Consulting, Shona Myers, Tim Lovegrove, Kate Lewis, Sian Carvell, Craig Pauling, John Stewart, Catherine Stephenson, Paul Mannix, Earl Shaver, Mike Frankcombe, Alastair Rylatt, Ed Bernacki and more. Thanks also to Johann Bernhardt and Eddie van Uden for allowing me to tell the story of how they created a test that everyone will pass, and to the many other people who allowed me to tell their stories. Thanks to Anthony Vallyon for the quote 'GDP – 1/All Other Indicators – 0' – brilliant! Thanks too to Lisa Martin for revealing to me the magic of the six capitals and the Global Reporting Initiative

(GRI), and to Maggie Tarver, who took a group of hopeful New Zealander authors to the 2012 Frankfurt Book Fair, where many of us found publishers.

I'm extremely grateful to my heroic clients, friends and colleagues who agreed to read this second edition in manuscript: a big thank you to Annette Lees, Beryl Oldham, Ghida Sinawi, Graeme Ridley, Dr Niki Harré, and my friend and business partner Susie Wood.

I have learned a great deal from the many expert trainers who generously present every month at meetings of the New Zealand Association of Training and Development – thanks to you all.

As always, thanks to my kind, knowledgeable and inspirational mentors, Ann Andrews CSP and Kim and Andrew Baird of Amazing Business.

Warm thanks to my publishers, Matthew Flynn and Anke Ueberberg of Gosbrook, for their faith in my topic and my book – it's been great to work with you!

And, of course, thanks to all the trainees who have attended my workshops over the years and from whom I have learned such a lot. No trainees, no training!

How to use this book

This is a 'how to' manual. Use it to get the big picture sorted. To help you with details such as standard forms, checklists and letters, you can access lots of free resources online.

ACTION PLANNER

Download the free Action Planner that accompanies this book from www.ESST. institute/Success/ActionPlanner or www.gosbrook.com/how-to-change-the-world. It contains action sheets to prompt discussion, research, reflection and action for your environmental training programme. The Action Planner starts with a simple spider diagram to help you to prioritize where you invest your effort.

As you read through the book, you will see prompts that encourage you to use the action sheets to apply your learning. They will look something like this:

 ACTION PLANNER

Where and how can you grow your green jobs? Use Action Sheet 1.1 in the free Action Planner that accompanies this book to explore how Storm Cunningham's 'eight great restoration industries' might relate to your environmental training priorities.

 TOOLBOX

Go to www.ESST.institute/Success/Toolbox to download lots of standard forms and letters, plus simple checklists, which will save you time as you set up or enhance your own environmental training programme.

1

Why there's never been a better time for environmental training

> Some will go and some will grow, but every job will change as we transition into the carbon neutral, circular economy.
>
> Samantha Smith, Just Transition Centre,
> International Trade Union Confederation (ITUC)[1]

1.1 A CLIMATE OF HOPE: A 'JUST' TRANSITION TO A SUSTAINABLE ECONOMY

> A qualified and well-informed workforce is the key to ensure the industry's responsible use of our planet's resources.
>
> UNESCO-UNEVOC[2]

There's never been a better time to do environmental training.

Why? Because there is a rapidly growing body of evidence that 'green' jobs can boost employment at the same time as improving social, economic and environmental outcomes.

And this is what people want: around the world, communities are voting and protesting on social and environmental grounds, while schoolchildren are shaming their elected representatives for 27 years of ineffectual action on climate change since the Rio Earth Summit in June 1992.[3]

In the six years since the first edition of this book, what's changed? Environmental issues are spreading and intensifying; we are already experiencing serious consequences of climate change; warnings from scientists, economists and insurers are becoming stronger; and many governments' actions still seem too slow and tentative.

But some positive movements have also emerged and they're picking up speed.

- **Business leadership** for the environment is stronger, better coordinated, more skilled and more clearly aligned with international initiatives such as the United Nations Sustainable Development Goals (SDGs).[4]

- **Financial agencies**, including the World Economic Forum (WEF), and the insurance sector are producing sophisticated analyses of the financial risks and costs of environmental harms, signalling that public and private infrastructure and other property already at risk from storm surges and sea level rise is no longer insurable.[5]

- **Investment interests** have, for many years, been documenting the higher returns from firms with better ethical and environment, social and governance (ESG) performance than their same-sector peers, while nearly 80 stock exchanges around the world now require their listed companies to prepare integrated annual reports that demonstrate the value they add to six forms of capital, including social and natural capital.

- **Governments** are realizing they must take prompt and effective action on climate change and supplement the use of gross domestic product (GDP) – which measures only the churn of money through the economy – with other indicators of success. Many governments now monitor a range of well-being indicators, including direct measures of happiness,[6] and several are developing well-being budgets that align the policy outcomes of different ministries and departments to more effectively deliver social, environmental and economic well-being.[7]

- **Workers' organizations** are building on research that came out of the 2008 global recession when the International Labour Organization (ILO) and the United Nations Environment Programme (UNEP) identified that, globally, the transition to a 'green economy' could yield 15–60 million jobs by 2032, lifting tens of millions of workers out of poverty while improving social and environmental outcomes.[8] The report said that 'the growth model of the past few decades has been inefficient, not only economically, but also from environmental, employment and social perspectives. It overuses natural resources, is environmentally unsustainable and has failed to meet the aspirations of a large proportion of society seeking productive, decent work and dignified lives'.[9] Such evidence has encouraged government well-being initiatives in many parts of the world.

- **Communities** are acting to protect their local environments and are creating local jobs and business opportunities by setting up initiatives to improve water quality, reduce landfill waste, restore vegetation and more. Such initiatives are common in both more and less developed countries. One outstanding example is the massive reduction of plastic bags and packaging from the Rwandan economy resulting from a grassroots movement supported by strong government measures, which created numerous new green jobs.[10] Where progress towards such goals appears slow, we are also seeing heightened environmental activism and global protests coordinated via social media, such as Extinction Rebellion.

- **Entrepreneurs** are setting up social and benefit enterprises to address environmental and social matters of interest and concern to them.[11] Whether they are not-for-profit or for-profit, they apply a commercial model to make a difference to their chosen well-being outcomes.

Responsible use of our plant's resources: what it's not – job losses; what it is – a shift in the focus of economic activity that creates *more* jobs that are good for people and places.

A 'new development model – one which puts people, fairness and the planet at the core of policy-making – is urgently needed, and is eminently achievable', according to the ILO.[12] Not only is it achievable, but also it's happening already. Storm Cunningham says that restoration of built and natural environments already constitutes a major, but overlooked, part of global economic activity and will soon account for the vast majority of development.[13]

And the economic need is great. Ecosystem services are the good things such as healthy air, water and soil, pollination and food, to name but a few, that the natural world provides for us for free. In 1997, Robert Costanza and others estimated that the value of ecosystem services to business equated to at least US\$33 trillion a year,[14] while a 2008 study estimated the annual economic cost of loss of ecosystem services by biodiversity and ecosystem degradation at 3.3–7.5 per cent of global GDP (around US\$2–4.5 trillion).[15] In 2014, Costanza and his colleagues revalued the monetary benefits of ecosystem services at US\$125 trillion a year – while also estimating annual losses in the value of these services at US\$4.3–20.2 trillion.[16]

Green jobs can transform these avoidable economic losses into health, social, environmental and economic gains. As Peter Bakker, President of the World Business Council for Sustainable Development (WBCSD), says: 'Transitioning to the circular economy represents a \$4.5 trillion opportunity, and has the potential to catalyse the biggest social, economic and environmental changes since the First Industrial Revolution.'[17]

In Storm Cunningham's restoration economy, eight 'giant, fast-growing industries are renewing our natural and built environments' – and creating vibrant businesses as they revitalize communities. He says that these already comprise a large proportion of economic activity, and that much more future economic growth will be based on renewing our natural, built and socio-economic assets by:

- restoring our **natural environments** – ecosystems, watersheds, fisheries and farms; and

- restoring our **built environments** – brownfields, infrastructure, heritage and places affected by natural disaster or war.[18]

The emerging focus on adult vocational training as a positive force for jobs and the environment shows the way ahead.

But gradual change is not enough. We need rapid change – change that doesn't simply destroy livelihoods, but creates new sustainable jobs that build on the skills of the old unsustainable jobs.

Forward-thinking bodies are setting up 'Just Transition Units' to support this change from sunset to sunrise sectors – from unsustainable to green jobs. In 2016, the International Trade Union Confederation (ITUC) set up the Just Transition Centre in Brussels with partners from around the world. The Centre 'brings together workers and their unions, communities, businesses and governments in social dialogue' to secure the 'future and livelihoods of workers and their communities in the transition to a low-carbon economy'. Training opportunities will help guarantee 'decent jobs … and greater job security for all workers affected by global warming and climate change policies'.[19]

The New Zealand government's Just Transition Unit was formed in May 2017 as part of a wider strategy to ensure that the transition to a low-carbon economy provides decent jobs, social protection, more training opportunities and job security for people affected by climate change and other environmental policies.[20] In April 2018, the government announced that no further offshore oil and gas exploration permits would be granted,[21] and it allocated significant funding for an action plan to support a 'just transition to a clean energy future'.[22] In making the announcement, Prime Minister Jacinda Ardern said: 'We're striking the right balance for New Zealand – we're protecting existing industry, and protecting future generations from climate change.'

The fossil fuels sector is an obvious priority for change,[23] and every community will be able to identify other economic sectors to support, helping them to tackle other pressing environmental issues while achieving multiple other social and economic goals.

 ACTION PLANNER

Where and how can you grow your green jobs? Use Action Sheet 1.1 in the free Action Planner that accompanies this book to explore how Storm Cunningham's 'eight great restoration industries' might relate to your environmental training priorities.

1.2   GROWING GREEN JOBS

> My definition of a green-collar job is this: a family-supporting, career-track job that directly contributes to preserving or enhancing environmental quality.
>
> Van Jones, author of *The Green Collar Economy*[24]

Every job will be a green job.

The ILO says that green jobs summarize the 'transformation of economies, enterprises, workplaces and labour markets into a sustainable, low-carbon economy providing decent work'.[25] Its updated Training Guidebook defines green jobs as decent jobs that:

- reduce consumption of energy and raw materials;
- limit greenhouse gas emissions;
- minimize waste and pollution;
- protect and restore ecosystems; and
- contribute to adapting to climate change.[26]

People in those skilled green jobs will need the:

> … skills for sustainable development, including the required technical skills, knowledge and values … in terms of social, economic and environmental development. Sustainable development skills relate to all facets of the society, not only including renewable energy, reuse and recycling of waste, utilization level of resources, green housing and sustainable planning, but also including wider areas, such as commerce, tourism, hospitality, information technology and finance [and more].[27]

We need to identify where environmental training can help the workforce in sunset sectors through a just transition to a sustainable economy and where environmental training may create new green jobs to take their place in a sustainable economy.

The fiscal multiplier is a useful tool for measuring the return on investment in green jobs. *Forbes* magazine cited a US study that estimated that when public money from taxpayers and ratepayers is invested into environmental restoration projects, 10.4–39.7 green jobs are created for every US$1 million invested, compared with the oil and gas industry, which supports about 5.3 jobs per US$1 million invested.[28]

Sophisticated tools are emerging to assess the potential for green jobs, such as the ILO's Training Guidebook on how to measure and model social and employment outcomes of climate and sustainable development policies.[29] The Guidebook's tools help countries to foresee the opportunities and risks involved in reshaping their economic and labour policies, and to make investment decisions that maximize employment gains and promote a just transition to the green economy. The analysis gives excellent guidance as to which sectors may need environmental training.

 ACTION PLANNER

Where and what are your sunset industries – or do you work for one? What are the implications for environmental training needs and partnerships? How do we 'green' conventional roles and organizations? Use Action Sheet 1.2 in the free Action Planner that accompanies this book to generate some ideas.

1.3 THE GLOBAL SKILLS SHORTAGE

> The only thing worse than training your employees and having them leave is not training them and having them stay.
>
> Attributed to Henry Ford, founder of the Ford Motor Company

A perfect storm of skills gaps, a complex education and training landscape, retiring baby boomers and younger people seeking more meaningful work have left employers and employees all around the world baffled and frustrated. Governments, businesses, educators and trainers must work together to grow the green skills the world needs.

Globally, many sectors are finding it difficult to recruit and retain skilled staff[30] or to find the training needed to upskill their staff to the level required.[31] This is particularly true for construction, a major contributor to the economy of many countries, and it's true for environment and sustainability skills across this and many other sectors.

Capacity and capability

Often used interchangeably, the two terms *capacity* and *capability* reflect different, but related, things. Broadly speaking, both set a ceiling on what a given group of people can achieve:

- **capacity** defines a ceiling set by headcount, funding and access to other resources; and

- **capability** defines the ceiling set by knowledge, skills and experience.

The transition to a more sustainable economy will simultaneously need and create capacity and capability for sustainability. Communities, companies and government bodies are increasingly aware of the regional and national development opportunities this offers[32] – and good environmental training will be essential to achieve it.

> **Case study: a strategic approach to enhancing industry capacity and capability**
>
> The acute skills shortage facing the post-earthquake rebuild teams in Christchurch, New Zealand, encouraged the development of a tiered approach to training. A partnership between the rebuild consortium and an industry training organization set up a strategic plan to manage training, recruitment and employment to meet a range of skills gaps.[33] Key initiatives formed a training pipeline, with:
>
> - 8–12 weeks pre-employment training to help people new to work to get 'work-ready';

- on-the-job retraining for applicants with work experience in other sectors;
- higher-level technical and supervisor training as employees progressed; and
- an on-the-job training model to support learners at work, including an innovative assessment resource that focused on practical delivery and observation of skills.

This enabled a seamless approach to upskilling learners at all levels and created a clear pathway for employees, resulting in immediate productivity gains for employers. With completion of the immediate earthquake recovery work, this has now been superseded by a range of active recruitment and workplace training programmes that address the country's ongoing skills shortage in the civil construction sector.[34]

Environmental capability gaps

Respondents to a survey of more than 900 UK businesses agreed that 'environment and sustainability skills will be essential to … survive and stay competitive, helping businesses adapt towards a "circular economy", extracting the maximum value from materials and turning waste into resources'.[35] But while respondents recognized the gravity of the threats, they also said they lacked the necessary skills to face them head on and turn them to their advantage. Moreover, gaining those skills was extremely difficult:

- 48.5 per cent reported problems recruiting environmental and sustainability professionals with the skills they need, and often had to recruit people with skills gaps;

- 62 per cent said that environment and sustainability training was more poorly funded than training in other professional areas, such as safety and finance, while 10 per cent had no environmental training budget at all; and

- 42 per cent had encountered barriers in securing environment/ sustainability training because of cost, time and lack of appropriate or local courses.

Capacity, demographic and recruitment gaps

There is also a wider gap: a global shortage of engineers, who are essential to many environmental initiatives and innovations. Global and national surveys have consistently found that supply can't meet demand for skilled engineers, with 82 per cent of Australian businesses responding to a 2008 survey reporting moderate or severe consequences to their business.[36]

To compound the problem, many sectors are experiencing a shortage of staff as baby boomers reach retirement age and not enough younger people are being attracted into key professions to replace them. Innovative workforce solutions can overcome this, for example people not wanting to fully retire are often happy to work part-time as mentors to new and young staff – and to be mentored in return on millennials' own skills.

Case study: capturing the expertise of the retiring baby boomers

Some years ago, the operations manager of a major utility company told me how he decided that, as a whole tranche of baby boomers approached retirement, he would film them talking about their routine procedures so that they could pass on their skills to the more junior people succeeding them. He got the baby boomers talking about anything and everything they could think of about the network and the different jobs they did to keep it in good shape. I loved the example he gave of one man describing the exact sound and vibration a valve makes as it's being properly bled – priceless value!

How simple would it be for you too to film such great knowledge and put it on your intranet or an open or closed channel on YouTube?

Many sector organizations are changing their recruitment methods to increase the diversity of their intake by, for example, offering exciting on-the-job learning opportunities, showcasing environmental and other experts, and encouraging young women and other groups into work, at all stages along the pipeline from schools through tertiary learning institutions to workplace support.[37]

Case study: a thank you to a community may reap rewards in the future

Big construction jobs can cause local inconvenience, but one company turned bad news into good news. Environmental manager Kylie Eltham was working on a major motorway project and, while the project finished ahead of schedule, it was unavoidably disruptive to local communities while it lasted. As part of its social responsibility programme, the company got together with local schools and organized field trips to show pupils the project and explain the environmental controls. The 8–10-year-old children were so intrigued that they came up with the idea of saying 'thank you' to the project team by painting the team's storage and flocculation sheds. The brightly coloured murals livened up the site for locals and the staff got a chuckle every time they saw them. And who knows how many of those children will go on to work on such sites?

Why people leave their jobs: the search for meaningful work

Surveys find that an inability to grow professionally is a leading reason why workers plan to leave their jobs. A 2019 report found that 28 per cent of millennials and 27 per cent of Gen Zs plan to leave their current organizations in the next two years because of a lack of learning and development opportunities.[38]

Providing poor training doesn't help: 41 per cent of employees at companies with poor training programmes plan to leave within a year, compared with 12 per cent of employees at companies that provide excellent training and professional development programmes.[39] When firms struggle to retain good staff, it is all the more important in a skills-short world for them to provide training to grow their own specialist skills in-house.

Moreover, young people increasingly seek meaningful work: as Punit Renjen, chief executive officer (CEO) of Deloitte, says: 'Millennials want to work for organizations that prioritize purpose as well as profit. It's as simple as that.'[40] One study found that, in 2018, only 48 per cent of millennials believed that businesses behave ethically, compared to 65 per cent in 2017.[41] Millennials don't 'see a clear vision from today's executives on how they will help improve society. These factors impact where millennials want to work and the roles they want to pursue. Companies need to offer what millennials want as well as what they need to succeed.'[42]

A 2019 Deloitte survey of 164,250 millennials and Gen Zs in ten countries found that nearly half want to be able to make positive impacts on their community/society, while a third said that climate change/protecting the environment topped the list of 20 challenges facing society that most concerned them.[43] A perceived 'profit over planet' attitude in business was out of step with millennials' priorities: 27 per cent believed that business should try to improve and protect the environment, but only 12 per cent thought that business was actually achieving this. The authors concluded by saying that the survey 'provides compelling clues for business leaders to follow': 'Millennials and Gen Zs show deeper loyalty to employers who boldly tackle the issues that resonate with them most, such as protecting the environment.'[44]

My 25 years of environmental training have shown me that those companies that develop their staff skills reap the rewards in terms of staff retention, engagement and productivity. Research confirms this.[45]

If we can't recruit the skills we need, we have to grow them. That's what this book is for.

 ACTION PLANNER

What statistics do you have about workforce demographics and trends in your sector of interest? What capacity and capability issues do you see? What are the implications for recruitment, retention and training of people with environmental skills? Use Action Sheet 1.3 in the free Action Planner that accompanies this book to consider these ideas.

1.4   GREEN REGULATION GROWS JOBS EVEN IN TOUGH TIMES

> If you want creativity, take a zero off your budget. If you want sustainability, take off two zeros.
>
> Jaime Lerner, celebrated urbanist and former mayor of Curitiba[46]

Going green won't kill jobs during hard times *or* when times are good.

Who would have thought that ordinary human beings could have favourite economists? These days it seems we all do, as we see leaders such as Herman E. Daly, Lord Nicholas Stern, Judge Professor Mervyn E. King, Robert Costanza, Angus Deaton, Amartya Sen, Richard Layard, Thomas Piketty and many more embracing the sustainability challenge.

Other leading economists, including Nobel prize-winners such as Joseph Stiglitz, Elinor Ostrom, Angus Deaton and William Norshaus, are bringing climate change, the commons and happiness into mainstream economic thinking.[47] Now, we must bring what have been previously treated as social and environmental externalities – that is, things that are not counted – into the heart of our economic analyses. A new approach to macro-economics for sustainability must provide for the things that make life worthwhile. Happiness or well-being economics has been on the rise for many years.[48] Economics for sustainability, or the green economy, including happiness, can also be a driver of economic development, partly through its beneficial effects on innovation.[49]

Today's governments have many policy levers to improve well-being across economic, social and environmental indicators. But can we make the necessary changes to a more sustainable world using the soft tools of moral suasion, soft policy and training? No, not alone.

Regulation also has a key role to play and we have hard evidence that it doesn't destroy jobs. Macro-economist Josh Bivens investigated the employment effects of the December 2011 US law approving environmental regulations to reduce emissions of mercury, arsenic and other toxic metals.[50] It was set to prevent up to 11,000 premature deaths each year and deliver many other health benefits, but, pre-passage, a lot of people were concerned it would 'kill jobs'. When Bivens investigated it in detail, he found that, far from killing jobs, the 'toxics rule' could create more than 100,000 jobs in the US by 2015.[51]

Bivens' message is that 'going green won't kill jobs during hard times': when the economy is doing well, environmental regulation has no effect on job growth, but when it isn't, such regulation is very likely to create jobs. These days, we need more jobs – and we need green jobs most of all.

There is a skills gap here – and environmental training can bridge it.

1.5   ROAD MAP: WHERE THIS BOOK WILL TAKE YOU

> Technical and vocational education and training (TVET) is central to
> facilitating the transition to green economies and sustainable societies
> and achieving the international community's ambitious 2030 Agenda for
> Sustainable Development and its 17 Sustainable Development Goals.
>
> UNESCO[52]

You can't manage what you don't measure. Or, as leading happiness economists put it, 'if you don't measure it, you don't treasure it'.[53]

This book focuses on how to set up successful environmental training programmes. However, along with its targeted environmental benefits, environmental training also delivers a wide range of associated economic and social benefits. We need to measure these if we are to track our progress towards the full suite of outcomes that define sustainability.

My planetary training model is a radically different approach to evaluating the effectiveness of environment and sustainability training. It aligns the measurable and monetized outcomes of environment and sustainability training with how governments, stock exchanges, companies and not-for-profits measure the outcomes of their activities.

In this book, I'll show you how to evaluate the effectiveness of your environmental training and align it with globally accepted social, economic and environmental indicators used by businesses, scientists and governments around the world.

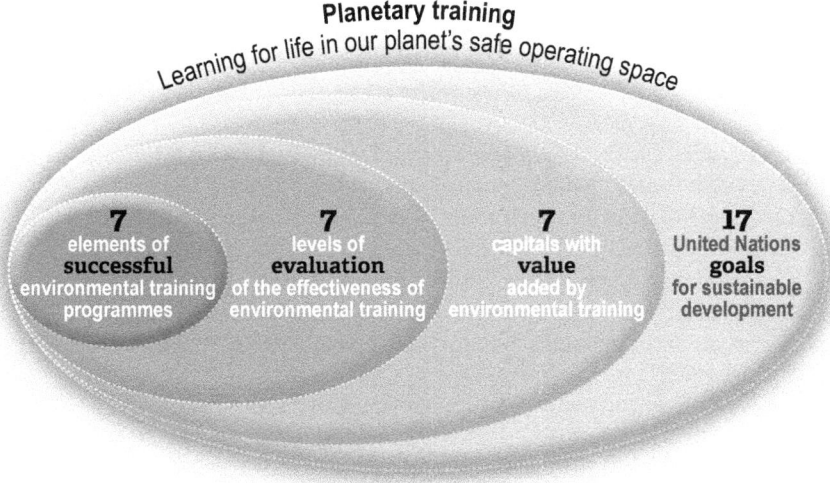

Figure 1.1 Planetary training: learning for life in our planet's safe operating space

In Figure 1.1, the two left-hand ellipses are training-related, so this book covers them in some detail. I then indicate how you can link your evaluation results to the other ellipses, which are more broadly sustainability-related:

- seven elements of successful environmental training programmes;

- seven levels of evaluation of the effectiveness of training;

- seven capitals across which successful environmental training adds value, including the six capitals on which growing numbers of stock exchanges require their listed companies to report;[54] and

- the 17 United Nations Sustainable Development Goals (SDGs).[55]

These and similar monitoring frameworks track our progress towards living within the biogeochemical planetary boundaries proposed in 2009 by Johann Rockström of the Stockholm Resilience Center and an international team. Surrounding these frameworks in Figure 1.1 is the safe operating space they defined in which humanity can continue to develop and thrive.[56]

Let's look next at just what 'training' is.

NOTES

1 An interview with Samantha Smith, director of the Just Transition Centre, broadcast on Radio New Zealand on 4 October 2018. Available at www.rnz.co.nz/national/programmes/ninetonoon/audio/2018665317/how-to-achieve-just-transition-to-green-jobs [accessed 4 October 2018].

2 UNESCO-UNEVOC is the United Nations Educational, Scientific and Cultural Organization's specialized Centre for Technical and Vocational Education and Training (TVET). It assists UNESCO's 195 member states to strengthen and upgrade their TVET systems. Find out more at www.unevoc.unesco.org/go.php [accessed 7 April 2019].

3 See e.g. the speech by then 12-year-old Canadian Severn Suzuki at www.youtube.com/watch?v=oJJGuIZVfLM (see a transcript at www.permacultureproject.com/from-the-rio-summit-a-speech-by-twelve-year-old-severn-suzuki/) and her speech to heads of state 20 years later at Rio+20 in 2012 at www.youtube.com/watch?v=1FmSxmpitBA; see also 15-year-old Greta Thunberg's speech in December 2018 to the UN Climate Change COP24 Conference in Katowice, Poland, at COP24 at www.youtube.com/watch?v=VFkQSGyeCWg [all accessed 17 March 2019].

4 See www.un.org/sustainabledevelopment/ [accessed 7 April 2019].

5 See e.g. World Economic Forum (2019) The global risks report 2019. Available at www.weforum.org/reports/the-global-risks-report-2019; UNEP Finance Initiative's Principles for Sustainable Insurance at www.unepfi.org/psi/the-principles/; Insurance Council of New Zealand (2019) LGNZ sea-level rise report

not even half the story. Media release, 1 February. Available at www.icnz.org. nz/media-resources/media-releases/single/item/ gnz-sea-level-rise-report-not-even-half-the-story/; Mina Martin (2017) LGNZ and ICNZ on the threat to coastal properties. *Insurance Business NZ*, 23 January. Available at www. insurancebusinessmag.com/nz/news/breaking-news/lgnz-and-icnz-on-the-threat-to-coastal-properties-52857.aspx; LGNZ (2019) $14 billion of council infrastructure at risk from sea level rise. Media release, 31 January. Available at www.lgnz.co.nz/news-and-media/2019-media-releases/14-billion-of-council-infrastructure-at-risk-from-sea-level-rise/ [all accessed 7 April 2019].

6 Global Happiness Council (2019) *Global happiness and well-being policy report.* Presented to the World Government Summit in Dubai on 10 February. Available at www.happinesscouncil.org/report/2019/ [accessed 7 April 2019].

7 Clare Feeney (2018) One number to rule them all: are we stuck with GDP? *Pure Advantage*, 31 July. Available at www.pureadvantage.org/news/2018/07/31/one-number-to-rule-them-all/ [accessed 7 April 2019].

8 Green Jobs Initiative (2012) *Working towards sustainable development: Opportunities for decent work and social inclusion in a green economy.* A joint ILO/UNEP study published on 12 June. Available at www.ilo.org/global/publications/ilo-bookstore/order-online/books/WCMS_181836/lang--en/index.htm [accessed 5 June 2019].

9 International Labour Organization (2019) *Working towards sustainable development: Opportunities for decent work and social inclusion in a green economy.* Available at www.ilo.org/global/publications/ilo-bookstore/order-online/books/WCMS_181836/lang--en/index.htm [accessed 16 September 2019].

10 Listen, from 29m30s, to the remarkable and inspiring interview at www.radionz. co.nz/national/programmes/saturday/audio/2018689844/colin-keating-25th-anniversary-of-the-rwandan-genocide; see also the nation-by-nation report at www.breakfreefromplastic.org/wp-content/uploads/2018/04/Stemming-the-plastic-flood-report.pdf and the paper 'Rwanda as role model' at www.mari-odu. org/academics/2018s_adaptation/commons/library/Otger1232018.pdf [all accessed 7 April 2019].

11 See the definitions at www.investopedia.com/terms/s/social-enterprise.asp and https://en.wikipedia.org/wiki/Social_enterprise, as well as the investment opportunities at https://impakter.com/ [all accessed 7 April 2019].

12 International Labour Organization (2019) *Working towards sustainable development: Opportunities for decent work and social inclusion in a green economy.* Available at www.ilo.org/global/publications/ilo-bookstore/order-online/books/WCMS_181836/lang--en/index.htm [accessed 16 September 2019].

13 Storm Cunningham (2002) *The restoration economy: The greatest new growth frontier.* Berrett-Koehler. Find out more at www.stormcunningham.com/ [accessed 29 April 2019].

14 Millennium Ecosystem Assessment (2005) *Ecosystems and human well-being: Opportunities and challenges for business and industry.* Available at www.millenniumassessment.org/documents/document.353.aspx.pdf [accessed 5 June 2019].

15 Robert Costanza, Ralph d'Arge, Rudolf de Groot, Stephen Farber, Monica Grasso, Bruce Hannon, Karin Limburg, Shahid Naeem, Robert V. O'Neill, Jose Paruelo, Robert G. Raskin, Paul Sutton and Marjan van den Belt (1997) The value of the world's ecosystem services and natural capital. *Nature*, 387, p. 256, table 2. Available at www.biodiversity.ru/programs/ecoservices/library/common/doc/Costanza_1997.pdf [accessed 5 June 2019].

16 Robert Costanza, Rudolf de Groot, Paul Sutton, Sander van der Ploeg, Sharolyn J. Anderson, Ida Kubiszewski, Stephen Farber and R. Kerry Turner (2014) Changes in the global value of ecosystem services. *Global Environmental Change*, 26, pp. 152–158. Available at http://community-wealth.org/sites/clone.community-wealth.org/files/downloads/article-costanza-et-al.pdf [accessed 8 April 2019].

17 World Business Council for Sustainable Development (2018) *Driving a circular vision: Sims Metal Management joins the World Business Council for Sustainable Development*. Available at www.wbcsd.org/Overview/News-Insights/General/News/Sims-Metal-Management-Joins-the-World-Business-Council-for-Sustainable-Development [accessed 2 June 2019].

18 Storm Cunningham (2002) *The restoration economy: The greatest new growth frontier*. Berrett-Koehler. Find out more at www.stormcunningham.com/ [accessed 29 April 2019].

19 Find out more at www.ituc-csi.org/just-transition-centre and www.ituc-csi.org/climate-justice-and-industrial [both accessed 8 April 2019].

20 You can find the New Zealand Just Transition Unit at www.mbie.govt.nz/business-and-employment/economic-development/just-transition/ [accessed 8 April 2019].

21 See the government media release at www.beehive.govt.nz/release/planning-future-no-new-offshore-oil-and-gas-exploration-permits [accessed 14 April 2019].

22 Find out more about the Tapuae Roa Action Plan at www.makeway.co.nz/news/launch-of-bold-ambitious-roadmap-for-taranaki-s-economy/ [accessed 14 April 2019].

23 Robin Martin (2019) Beyond gas and oil: can alternative energy save Taranaki? A radio documentary and article broadcast on Radio New Zealand's *Insight* on 2 June. Available at www.rnz.co.nz/national/programmes/insight/audio/2018696911/beyond-gas-and-oil-can-alternative-energy-save-taranaki [accessed 2 June 2019].

24 Van Jones (2008) *The green collar economy: How one solution can fix our two biggest problems*. HarperOne, p. 17.

25 See www.ilo.org/global/topics/green-jobs/lang--en/index.htm [accessed 5 June 2019].

26 Green Jobs Assessment Institutions Network (GAIN) (2017) *How to measure and model social and employment outcomes of climate and sustainable development policies: Training guidebook*. International Labour Organization. Available at www.ilo.org/global/topics/green-jobs/areas-of-work/gain/training-guidebook/lang--en/index.htm] [accessed 12 April 2019].

27 The October 2012 China-Australia Green Skills Conference was themed *Green Skills: Powering a Better Future*. The extract was available at http://en.ceaie.edu.cn/en_declare.php and http://en.ceaie.edu.cn/en_news_detail.php?id=5377 in November 2012, but the pages are no longer live.

28 Logan Yonavjak (2014) Now THIS is what we call green jobs: the restoration industry 'restores' the environment and the economy. *Forbes*, 8 January. Available at www.forbes.com/sites/ashoka/2014/01/08/now-this-is-what-we-call-green-jobs-the-restoration-industry-restores-the-environment-and-the-economy/; see also K. Todd, T. BenDor, W. Lester, A. Livengood, A. Davis and L. Yonavjak (2013) *Exploring and understanding the restoration economy*. Available at https://curs.unc.edu/files/2014/01/RestorationEconomy.pdf [both accessed 29 April 2019].

29 Green Jobs Assessment Institutions Network (GAIN) (2017) *How to measure and model social and employment outcomes of climate and sustainable development policies: Training guidebook*. International Labour Organization. Available at www.ilo.org/global/topics/green-jobs/areas-of-work/gain/training-guidebook/lang--en/index.htm [accessed 12 April 2019].

30 Susan Freeman-Greene, CEO of Engineering New Zealand, in an interview with Jesse Mulligan broadcast on Radio New Zealand on 14 September 2018. Available at www.radionz.co.nz/national/programmes/afternoons/audio/2018662565/engineering-in-new-zealand [accessed 2 June 2019].

31 BDO (2018) *The construction survey report 2018*. Available at www.bdo.nz/en-nz/industries/construction-and-real-estate/2018-construction-survey-results-en [accessed 16 August 2019].

32 Robin Martin (2019) Beyond gas and oil: can alternative energy save Taranaki? A radio documentary and article broadcast on Radio New Zealand's *Insight* on 2 June. Available at www.rnz.co.nz/national/programmes/insight/audio/2018696911/beyond-gas-and-oil-can-alternative-energy-save-taranaki [accessed 2 June 2019].

33 The InfraTrain and SCIRT training partnership is no longer accessible online. You can find out more about the SCIRT legacy at https://scirtlearninglegacy.org.nz/ [accessed 2 June 2019].

34 See e.g. the successful EP!C programme, developed by Civil Contractors New Zealand to 'show people the gateway to working in an industry they can be proud of'. Find out more at https://epicwork.nz/ and https://civilcontractors.co.nz/professional-development/epic-careers-in-infrastructure/; see also the Civil Trades Certification and Recognition of Current Competence programme at www.connexis.org.nz/civil/civil-trades/ [all accessed 9 August 2109].

35 Institute of Environmental Management and Assessment (2014) *Preparing for the perfect storm: Skills for a sustainable economy*. IEMA. pp. 5–6. Available at https://cdn.ymaws.com/sustainabilityprofessionals.site-ym.com/resource/resmgr/documents/ss_documents/perfect_storm-_amended_versi.pdf [accessed 16 September 2019].

36 (1) The Queen Elizabeth Prize for Engineering (QEPrize) (2018) *Create the future report*. Available at https://qeprize.org/report/, summarized in Jonny Williamson (2018) Shortage of engineers could curb global economic growth. *The Manufacturer*, 26 March. Available at www.themanufacturer.com/articles/shortage-of-engineers-could-curb-global-economic-growth/ [both accessed 2 June 2019]. (2) Engineers Australia (2008) *The engineering profession: A statistical overview*, 5th edn, p. 74. (3) Teletrac Navman (2019) *Construction industry report 2019*. Commissioned by Teletrac Navman and Civil Contractors New Zealand (CCNZ), cited in James French (2019) Lack of certainty threatens job security. *Contractor*, August. Available at www.teletracnavman.co.nz/lp/gc/confirmation?survey_results [accessed 9 August 2019].

37 See the EP!C programme at https://epicwork.nz/ [accessed 9 August 2109].

38 Deloitte (2019) *The Deloitte global millennial survey 2019*. Available at www2.deloitte.com/global/en/pages/about-deloitte/articles/millennialsurvey.html [accessed 8 August 2019].

39 Emmanuel Olaoye and Stuart Gittleman (2014) Effective training a weak link in many compliance programs: survey. *Compliance Complete*, 13 August. Available at https://blogs.reuters.com/financial-regulatory-forum/2014/08/13/effective-training-a-weak-link-in-many-compliance-programs-survey/ [accessed 2 June 2019].

40 Punit Renjen (2016) Getting inside the heads of millennials. *LinkedIn*, 13 January. Available at www.linkedin.com/pulse/getting-inside-heads-millennials-punit-renjen/ [accessed 8 August 2019], cited in S. Shameem (2019) Things that millennials expect from their jobs. *Entrepreneur India*, 25 March. Available at www.entrepreneur.com/article/331111 [accessed 8 August 2019].

41 Joan Kuhl (2018) What do millennials want in their careers? *Forbes*, 23 July. Available at www.forbes.com/sites/joankuhl/2018/07/23/what-do-millennials-want-in-their-career/#65b56e66fcb1 [accessed 8 August 2019].

42 Ibid.

43 Deloitte (2019) *The Deloitte global millennial survey 2019*. Available at www2.deloitte.com/global/en/pages/about-deloitte/articles/millennialsurvey.html [accessed 8 August 2019].

44 Ibid., p. 26.

45 Dr Laurie Bassi (2000) *Profiting from learning: Do firms' investments in education and training pay off? Investing in training improves financial success*. An executive summary of research conducted in 2000 for the American Society for Training and Development (ASTD), now the Association for Talent Development (ATD). Available with permission from ASTD at https://businesstrainingexperts.com/knowledge-center/training-roi/profiting-from-learning/ [accessed 23 June 2019].

46 Roberta Brandes Gratz (2017) Curitiba's Jaime Lerner. *Huffington Post*, 6 December. Available at www.huffpost.com/entry/curitibas-jaimie-lerner_b_4179203 [accessed 15 August 2019].

47 Clare Feeney (2018) One number to rule them all: are we stuck with GDP? *Pure Advantage*, 31 July. Available at https://pureadvantage.org/news/2018/07/31/one-number-to-rule-them-all/ [accessed 22 April 2019]. See also the list of Nobel

laureates in economics at https://en.wikipedia.org/wiki/List_of_Nobel_Memorial_ Prize_laureates_in_Economics [accessed 21 July 2019].

48 C. Graham (2011) *Happiness economics: Can we have an economy of wellbeing?* Available at https://voxeu.org/article/happiness-economics-can-we-have-economy-wellbeing [accessed 12 April 2019]. Carol Graham is Leo Pasvolsky Senior at the Brookings Institution and College Park Professor in the School of Public Policy at the University of Maryland.

49 David A. Fleming-Muñoz and Stephan J. Goetz (2018) Happiness helps football players do better, and it could help economies too. *The Conversation*, 10 July. Available at www.theconversation.com/happiness-helps-football-players-do-better-and-it-could-help-economies-too-99286 [accessed 22 April 2019]. Fleming-Muñoz is an economist at the Australian CSIRO and Goetz is Professor of Agricultural and Regional Economics, Pennsylvania State University.

50 Josh Bivens (2012) Going green won't kill jobs during hard times. *New Scientist*, 24 March.

51 Josh Bivens (2012) *The 'Toxics Rule' and jobs: The job-creation potential of the EPA's new rule on toxic power-plant emissions*. Issue Brief #325 of the Economic Policy Institute, a non-partisan think tank in Washington DC, 17 February. Available at www.epi.org/publication/ib325-epa-toxics-rule-job-creation/ [accessed 5 June 2019].

52 UNESCO (2016) *Strategy for Technical and Vocational Education and Training (TVET) (2016–2021)*. UNESCO, pp. 4 and 6. Available at https://unesdoc.unesco.org/ ark:/48223/pf0000245239; see also https://sustainabledevelopment.un.org/ post2015/transformingourworld [both accessed 29 October 2019].

53 John F. Helliwell, Richard Layard and Jeffrey Sachs, eds (2015) *World Happiness Report 2015*. New York: Sustainable Development Solutions Network, p.118. Available at https://worldhappiness.report/ed/2015/ [accessed 20 October 2019]. More often quoted in the negative form cited, the original quote is 'If you treasure it, measure it.'

54 Find out more at https://integratedreporting.org/the-iirc-2/ [accessed 7 August 2019].

55 Find out more about the UN SDGs at www.un.org/sustainabledevelopment/ sustainable-development-goals/ [accessed 12 September 2019].

56 Find out more about this important work at www.stockholmresilience.org/ research/planetary-boundaries.html and www.stockholmresilience.org/research/ research-news/2019-07-25-its-all-about-the-safe-operating-space.html [both accessed 7 August 2019].

2

What exactly is 'training'?

Essentially, the definition of training is unlocking potential through tracked
and measured knowledge sharing.

Lessonly.com[1]

2.1 TRAINING DEFINED

Good organizations explore ways to recruit, retain, and train their
employees. The rest make it easy to poach their best talent.

Jessica Miller-Merrell, Workology.com[2]

**Successful training changes us as trainers as much as – or more than – it changes our
trainees.**

The words 'training', 'learning', 'capability', 'awareness' and 'education' are often
used interchangeably. Other terms like 'professional development' or 'training
and development' are also common. In this book, I use the term 'training' in a
very specific way.

Training is the acquisition of work-related knowledge, skills and practices that will improve
a specified aspect of on-the-job performance in objectively observable ways, as defined in
a clear statement of performance standards and/or outcomes.

This book focuses on environment and sustainability performance in the workplace. Because this is an area that can be subject to environmental law, with
consequences of breach including fines and imprisonment, a core principle is
that of fairness: training must ensure that trainees know exactly what actions are
expected of them to meet their compliance requirements.

Throughout this book, therefore, my use of the term 'training' implies the existence of measurable performance standards, benchmarks or outcomes specified
by an appropriate authority such as an environmental regulator, or industry or
professional association or similar body, in terms that enable the defined environmental practices to be:

- performed to the required standard by the people who must meet it;

- transparently and consistently assessed against that standard by the specifier or their agents; and

- supported by prompts and management systems that allow the new learning to be practised and improved in the workplace – because it's no use sending staff to training workshops if their workplace does not provide the encouragement, time, budget and resources they need to put what they have learned to good effect.

If you don't have a guideline or standard in place, don't worry: we'll cover the topic in Chapter 3. Not comfortable with compliance? Don't panic – Chapter 3 looks at how to do it well and fairly.

Good training defines the required workplace competencies and may or may not be supported by other statutory and non-statutory measures to encourage, require or enforce the performance standard.

Great training can also be so much more than this. The truth of the old adage 'the best way to learn is to teach' has been demonstrated by robust research.[3] Successful training transforms trainees and trainers into lifelong learners inspired by the joy of learning.

As we move into an era of green jobs, we will be creating environmental jobs no one has ever heard of before. But reassure yourself as you move into this new territory: you don't have to be an expert to start with. Just start learning and mixing with others doing the same thing, and your expertise will grow over time.

Most of the people I know who have set up successful environmental training programmes had no idea what they were getting into when they started. The team who set up the major training programme on urban erosion and sediment control that first enthralled me were experts at rural soil conservation – but they had to learn about urban soil conservation on big, fast-moving and temporary construction sites, a field in which they were novices. They understood soil and water, and, in writing their guidelines, they learned a great deal about erosion and sediment control on construction sites. But it took probably five or six years before people working with those sites really became experts in what was then a new field in their country – and they're still learning. *Real experts never stop learning!*

Figure 2.1 shows how becoming a genuine expert is a personal, as well as professional, journey of lifelong – and, as Jost Reischmann says, lifewide[4] – learning.

 ACTION PLANNER

Want to get more clarity on the difference between training and other commonly used terms? Use Action Sheet 2.1 in the free Action Planner that accompanies this book.

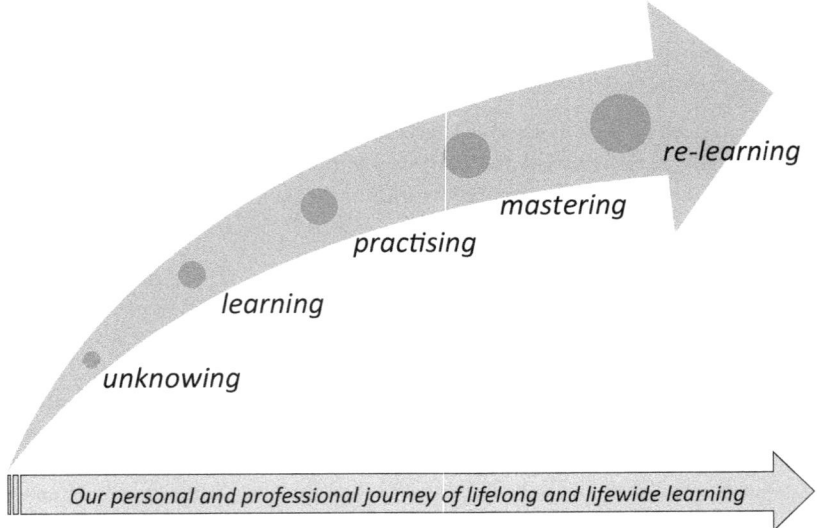

Figure 2.1 Becoming an expert in a new green job

2.2 WHAT OTHER EDUCATION AND TRAINING IS OUT THERE?

> Each time we share information and refine our techniques, we collectively become a little wiser and move a step closer to the sustainable future our children deserve.
>
> Dr Doug McKenzie-Mohr[5]

Show you know who else is out there delivering education or training on your issue of concern.

Your business case should identify existing providers of education or training to your sector. I've defined 'training' at the start of this chapter, so here is my definition of 'education'.

Education is the process of receiving or giving systematic instruction, especially at a school, technical college or university.

The following is an overview of the education and training landscape common across most countries.

- **Formal education**   This type of education and training offers recognized qualifications, including through:
    - universities; and
    - technical colleges or similar institutes and other public or private training providers authorized to deliver training and confer

qualifications in the technical and vocational education and training (TVET) space. These can include industry skills bodies that focus on meeting the needs of industry sectors such as construction, infrastructure, solid waste handling and the like.

- **Formal recognition of skills and experience** People working in skilled trades can be awarded qualifications by going through a formal process that recognizes their current competence rather than requiring them to go back to training or apprenticeship. For semi-professionals and people who have existing qualifications, there is growing interest in digital micro-credentials, which allow working people to submit examples of their skills for assessment and to be awarded a qualification from the relevant vocational training body.

- **Formal professional training** This is formally recognized by professional bodies such as those for engineers, planners, architects, surveyors, trainers and other professions. The recognition of this training is through systems such as professional registration processes and annual requirements for continuing professional development (CPD). These bodies often also offer training on a wide range of topics and may issue various forms of recognition for this.

- **Other recognition of training** Additionally, commercial or not-for-profit training providers may:
  - issue their own certification that licenses people to carry out certain tasks (common for health and safety and hazardous substances training); and/or
  - issue attendance certificates for attending a public or in-house training workshop.

- **In-house training providers** These will often give some form of recognition to the organization's trainees and larger firms will usually track all training delivery using the human resources (HR) department's learning management system (LMS).

University courses in engineering, geography, environmental science, biology, ecology, landscape design, planning, architecture, surveying and the like produce graduates who are good generalists or good specialists (we need both). Such courses keep up or ahead of new developments in their field, but specific environmental workplace practices are not normally included.

Some trades training delivered by technical colleges contains environmental information, and TVET providers are often very willing to work with government and business stakeholders to develop new courses of training on environmental topics.

Developing countries face particular sustainability issues, and the United Nations Institute for Training and Research (UNITAR) provides training, learning and knowledge-sharing events focused on supporting their governments to achieve the United Nations Sustainable Development Goals (SDGs).[6] It has a wide range of face-to-face and online opportunities on offer and, in 2017, it delivered 497 individual activities benefiting more than 56,000 participants. UNITAR now provides learning solutions to institutions and individuals from both public and private sectors all around the world. This option is well worth exploring.

Many bodies provide third-party verification for a range of environmental standards, including those in the International Organization of Standardi (ISO) family. They effectively train their clients in carbon emissions reduction, infrastructure sustainability, lifecycle assessment, organic farming, energy efficiency, environmental management, and many other environment and sustainability topics. These clients then often upskill staff in their organizations. If these certification bodies don't deliver formal training on the topic of concern for you, it could be worth approaching them to help you.

Most in-house and public environmental training programmes, including those delivered by government bodies, target very specific skills needed by people with academic and trades training in a space roughly defined by formal training for the purposes of CPD. All components of the education and training landscape have a role to play in great environmental training, and there is great value in working with your tertiary providers, as we've seen.

It is the CPD spaces in Figure 2.2 that are the focus of this book. 'Continuous professional development' is another way of saying lifelong learning.

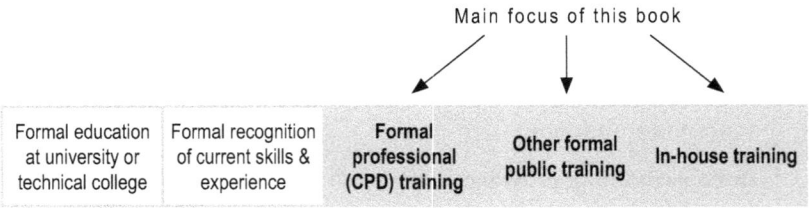

| Formal education at university or technical college | Formal recognition of current skills & experience | Formal professional (CPD) training | Other formal public training | In-house training |

Figure 2.2 Focus of this book

 ACTION PLANNER

Use Action Sheet 2.2 in the free Action Planner that accompanies this book to help you to locate environmental training delivered within or beyond your own organization.

2.3 WHO MOST NEEDS ENVIRONMENTAL TRAINING?

> Climate change is the latest urgent issue where lifelong adult learning and
> education has powerful potential.
>
> Shirley Walters, Emerita Professor of Adult and Continuing
> Education, University of Western Cape, South Africa[7]

Every sector of the economy needs environmental training – but not necessarily every person in every sector. Environmental training has traditionally targeted high-compliance sectors. It is now widening its scope to include sunset and sunrise sectors for a low-carbon economy.

In addition to the training providers summarized above, governments, not-for-profits and the business sector deliver environmental training on a wide range of topics.

Globally, it's reasonably common for government bodies to deliver environmental training to businesses. Worldwide, depending on local and national issues, common topics include:

- erosion and sediment control;

- pollution prevention, waste minimization and resource efficiency for manufacturers;

- construction and demolition waste minimization; and

- river restoration and replanting of streambanks and wetlands.

Not-for-profits deliver training on a wide range of issues. Biodiversity, biosecurity, and control of plant and animal pests are common topics in New Zealand, for example.

Increasingly, companies, especially in high-compliance sectors, also deliver their own, more detailed, in-house environmental training via programmes that become increasingly comprehensive for bigger firms with a wider array of different environment and sustainability risks. Commercial providers, including third-party verifiers, sometimes deliver training as part of their process, while topics such as reduction of greenhouse gas emissions, solid waste minimization and more are hot topics.

Here's how David Wentworth brilliantly defines high-compliance sectors:

- people can die;

- people can go to jail; and/or

- it costs a lot of money when you get it wrong.[8]

High-compliance firms include manufacturing, waste management, utilities and civil construction firms for which systems failure can cause immediate environmental harm and in which worksite staff need to be highly trained in prevention and response and constantly alert. They will increasingly include sunset industries, such as the fossil fuels, and the wider energy sector as emissions controls bite deeper.

Remembering my very strict definition, do we need 'training' to tell us how to switch off a light? Is environmental training, as I define it, needed in every workplace? Here's what I think.

- Yes, *every organization* – be it business, government or not-for-profit – needs to play an active role in accelerating the transition to a sustainable economy in which people and the environment really count.

- No, *not every organization* will need to formally train all its staff to help them to play their role in this exciting transition.

- But, yes, *every sector* will need trained people to help to roll out the changes in workplace practice that will get us there.

This means it's not always people doing the hands-on work who need environmental training. It may be the people working for a professional or trades association who need the training, so that they can work out what sustainability actions to disseminate to their members.

An example is the drive for more sustainable retail in the UK some years ago, which saw large retail companies recruiting entire sustainability teams: they were effectively creating a new workforce of sustainability professionals in the retail sector.[9] These people can undergo the technical training they need to set the strategic sustainability direction of their firms, and they can set up the systems to support their staff to carry out specific actions to reduce carbon emissions, plastics use, other solid waste, water and energy use, and so on, including norms and reminders about turning off lights and equipment, and sorting waste into compostables, reusables, recyclables and the like.

But what about the small retailers who may have only a handful of staff? They don't have the capacity to do all those things for their own tiny firms. Here's where the professional and trades associations come to the party: if they take the lead for the whole sector, they can disseminate a simple action plan to the small players and provide support that helps them to follow it.

Essentially, this is about whether environment and sustainability skills are best delivered by a more centralized or more distributed need for training across a sector, or more in-house vs more publicly available training. Understanding this will help you to work out who the main players are for your environmental training, so that you can make the most cost-effective use of your environmental experts.

 ACTION PLANNER

What high-compliance sectors are in your jurisdiction – or do you work for one? What information do you have about the sector's or your own environmental performance? What would be the role of sector organizations in this? Use Action Sheet 2.3 in the free Action Planner that accompanies this book to answer these questions.

2.4 INTRODUCING THE SUBJECT MATTER EXPERT

SMEs [subject matter experts] might be great at their jobs but they are not great trainers. Just because someone knows the subject well doesn't mean they can teach it.

Professional trainer Bill Cushard, Allonhill[10]

Training is a global profession, full of experts every bit as skilled as our global profession of environmental experts. Let's make use of their specialized skills to make the best use of our own.

'Subject matter expert' (SME) is what professional trainers call those people whom they ask to get involved in training because their technical knowledge is so valuable.

Environment and sustainability training is traditionally the domain of a range of SMEs, such as those who know all about stormwater modelling, water pollution control, ecological restoration, contaminated soil management, emissions reduction, solid waste minimization and more – much more.

Here's where we run into Problem #1: any professional trainer will tell you that most SMEs in any field don't know that there is such a profession as learning and development for adult vocational training. So they just get stuck in and deliver training themselves without framing it in the context of the wider internal and external aspects of the business. And environment and sustainability SMEs are no exception.

Problem #2 then rears its ugly head: environmental experts don't usually know that there is a thing called 'train the trainer' training that will help them develop and deliver their training in ways that optimize the learning of the many different groups of adult learners they will face.[11] In my experience, these experts often make a good job of their training – but why settle for good when access to excellence is readily accessible if you know where to look for it?

Even assuming that an environmental SME undertakes some trainer training, my experience tells me that they will still face Problem #3: how to rigorously evaluate how effective their training is. Yes, they may have had to prepare a business case for their training to justify the cost. But they can measure so much more on the plus side of the balance sheet if they learn how to use the tools of the professional trainer to measure the effectiveness and direct and indirect benefits of their training.

This book provides solutions to these three problems.

2.5   STRATEGY AND PROGRAMME: HOW ARE THEY DIFFERENT?

> We always overestimate the change that will occur in the next two years and underestimate the change that will occur in the next ten.
>
> Bill Gates[12]

An organizational environment and sustainability strategy may contain several very different topic-specific training programmes.

What's the difference between a training strategy and a training programme – and why is it important?

- A **training strategy** helps you provide the right type of training to the right people at the right time – that is, when they need the skills and knowledge to help to move the organization forward in line with its priorities.[13] It sets out an approach to training in a way that is fully aligned with the wider external context within which your organization operates, its internal vision, mission, purpose and goals, and the risks and opportunities associated with both of these. It then aligns the purpose and outcomes of training with them, based on best adult learning principles and practices, in a way that can demonstrate the value of investing in training and the effective use of the training budget. A training strategy may identify the need for training on several different environmental topics and should prioritize them.

- A **training programme** is specific to a single environment and sustainability topic. It delves into the detail of training needs and desired workplace performance, as well as trainees' learning needs and preferences. It develops and delivers training within a planned time frame, and it supports trainees and their supervisors and managers to enhance application of new learning in the workplace. A programme also sets out the indicators and methods with which to evaluate the specific outcomes of the training in terms of the trainees' desired performance, as well as the organizational and other outcomes in the training strategy.

You may find it helps your business case to include a PEST and SWOT process to develop a situational analysis and an organizational analysis. The PEST analysis is typically performed first and mainly defines those factors in the external context in which your environmental training programme will operate that affect you and your programme stakeholders. It looks at the:

- **Political/legal** context, such as more stringent environmental compliance standards and/or heavier enforcement;

- **Economic** context, such as carbon taxes or emission schemes, financial incentives and penalties;

- **Social/demographic** context, for example fair trade and responsible consumer movements and communities lobbying for better environmental standards; and

- **Technological** context, such as new and more sustainable processes.

Many people consider further factors, such as Environmental/Ecological, Ethical, Legal and International. A search for other PEST acronyms such as PESTEL or PESTLE will find them. You might want to consider these as well to make sure you cover all your areas of risk.

The SWOT analysis looks largely at internal factors that affect your programme and its stakeholders as individual agencies and in terms of your shared future for the sector of concern. These factors comprise:

- **Strengths**, such as existing environmental systems and programmes;

- **Weaknesses**, such as a record or culture of persistent environmental non-compliance or under-resourcing;

- **Opportunities**, including those opened up by new environmental regulation or partnerships with your stakeholders; and

- **Threats**, such as exposure to climate change and rising sea levels, shortages of water, energy or materials, and procurement or customer pressures.

Figure 2.3 maps the relationships between strategy and programme within the wider context in which your organization operates.

Figure 2.3 Training strategy and training programmes in context

If you need to deliver training on only a single environment or sustainability topic, then you can integrate your strategy and programme into one document.

This book focuses on training *programmes* – that is, how to develop, deliver and evaluate an effective training programme on a specific environment and sustainability topic.

 ACTION PLANNER

Does your training strategy focus on one topic or many? Will you need separate training programmes for very different training topics? If so, what criteria relating to risk and opportunity will you use to prioritize them? Use Action Sheet 2.4 and the solid waste example in the free Action Planner that accompanies this book to answer these questions.

Use Action Sheet 2.5 to apply the PEST and SWOT process to carry out a situational analysis and an organizational analysis.

2.6   THE SUCCESS STORY THAT INSPIRED THIS BOOK

> Ehara taku toa, he takitahi, he toa takitini. [Success is not the work of one, but the work of many.]
>
> Māori whakataukī (proverb)[14]

Identifying the elements of success allows us to replicate – and improve – them.

The unexpected success of a long-running environmental training programme in Auckland, New Zealand, led me to write this book. After some years of thought, I isolated the factors that made it so successful (the 'Success Framework' in the next chapter). I'd just written my second conference paper on environmental training, drawing on my experience with several very different programmes for a wide range of clients, when the councils in Auckland went through a far-reaching reorganization. I realized I needed to tell the story of Auckland's erosion and sediment control training programme in case the knowledge was lost with the inevitable staff dispersion that would – and did – follow.

When the programme first started, my co-trainer Brian Handyside and I thought we might deliver two or three years of training, then stop, because everyone would be trained. What actually happened was quite different: 15 years later, not only was demand for the training still strong, but also it was endorsed by major government agencies that required their service providers to attend. Moreover, this highly successful programme had – like the programmes we'd looked at when starting out – inspired a number of similar programmes. We even had people from other countries attend our workshops to find out what we were doing.

Now that the local government reforms have bedded down, the programme has been reborn, and the council is now delivering training against new technical guidelines.

But how did it all begin? The short answer is as follows.

New Zealand's biggest city, Auckland, straddles three of the country's biggest estuaries and two beautiful open coastlines on the West coast and also on the more vulnerable East coast, where development is focused.

 Rapid urban growth in the 1970s leads to unsightly plumes of sediment in streams and harbours, clogged streams and stormwater systems, causes localized flooding and leaves deposits of sticky yellow clay on popular bathing beaches.

 Early research in the 1980s highlights the risks and come up with some solutions, but there are no legal policies or 'teeth' in the contemporary statutes and regulations to ensure that the solutions are used.

 A new environmental law is passed in 1991, more research is done in partnership with the development industry, and council staff work with the industry to write a more comprehensive technical erosion and sediment control guideline.

 In 1992 the council forms a liaison group, including land developers, local councils, consultants, contractors, environmental groups and a representative of the local indigenous Māori tribes.

 Under the new legislation, the council is able to write an enforceable policy plan requiring all land developers to apply for an environmental permit and adhere to the new guideline, and works with the liaison group and the wider industry to do this.

 The legal permitting and new guideline are so novel that, in 1995, the council works with the liaison group to set up a training programme to help the development sector to use the new guidelines.

 On-site inspections prove very helpful in bringing both industry and council staff up to speed. Good operators welcome the phased-in use of enforcement for non-compliant operators who undercut them in the market.

 By the early 2000s, we realized that the training had become a vital part of Auckland's land development sector, creating a whole new profession of environmental managers on large construction sites and being taken up around the country.

Figure 2.4 Learning how to build a successful environmental training programme

Erosion and sediment control in Auckland: a potted history of a training programme

Looking back at the steps taken in building a successful programme, we might easily perceive an elegant sequence of planned actions. The path to Auckland's erosion and sediment control training programme is just such a case – but, in truth, the programme managers and partners simply did what needed to be done, and each unexpected twist revealed the next necessary step.

Figure 2.4 summarizes the steps that led to success in Auckland's programme. In 10–15 years, our training programme became so successful that we ended up creating a whole new profession of environmental managers on large construction sites, whose role added bottom-line value to the construction sector by: making sites safer and easier to work in wet weather; reducing water damage to completed works and avoiding the high cost of associated repairs; and reducing the time and money spent on legal wrangles. The programme had spread across the country, creating new professionals who are educated, dedicated and mobile, building and transferring knowledge and skills as they move freely between the development, consulting, contracting, council and community sectors. This profession is increasingly stepping up to its role in a vibrant restoration economy.

The ebooks listed in Further resources for this chapter (p.200) provide more detail about the training programmes discussed here.

2.7 SAME, BUT DIFFERENT: OTHER SUCCESSFUL TRAINING PROGRAMMES

> There are three kinds of men: ones that learn by reading, a few who learn
> by observation and the rest of them have to pee on the electric fence and
> find out for themselves.
>
> Source unknown[15]

Environmental training programmes share the same general principles across different legal and natural contexts and different sectors.

Although it was an urban erosion and sediment control training programme that inspired me to write this book, I've developed and delivered training on topics as diverse as household hazardous waste, septic tanks, riparian restoration, solid waste minimization, industrial water pollution prevention and integrated catchment (watershed) planning. In all cases, my government, business and not-for-profit clients considered that the training results more than repaid the costs of the training – a return on investment (ROI) that we now know could have been strengthened even further had we performed a formal analysis before and after the training!

Environmental training programmes take place in a wide range of legal, sector and natural contexts, but the general principles remain the same, as the following case studies will show, which are from a mix of bodies including business, government, education and indigenous peoples.

Throughout this book, I emphasize the importance of measuring the outcomes of training programmes. The case studies that follow include a mix of qualitative and quantitative indicators of success, ranging from winning an award to adding up the dollar value of time saved in legal enforcement.

The rest of the book – especially Chapter 6 – shows how you to select robust indicators of the effectiveness of your training and evaluate them to the highest standards acknowledged by professional trainers throughout the world, including measuring the full financial ROI. One of my aims in writing this book is to see every environmental training programme include robust evaluation of its success.

 ACTION PLANNER

What can you learn from comparing the examples that follow with each other? How could this inform your existing or proposed training programme(s)? What indicators would you use to measure the success of these programmes? Use Action Sheet 2.6 in the free Action Planner that accompanies this book to highlight the similarities and differences between the following case studies of successful environmental training programmes.

Training tripled their turnover ...

The very worst-case scenario for a civil construction company is when the knock-on effects of poor environmental performance affect not only its financial bottom-line performance, but also its ability to retain clients and attract new work – in effect, its ability to stay in business.

That's what happened to one privately owned firm. It had been prosecuted several times by different councils across the country and had attracted a host of lesser enforcement actions. In 2004, the company had a turnover of NZ\$100 million, 74 projects on the books, 17 environmental infringement notices and 23 per cent of its environmental controls non-compliant with the technical guidelines. The chief executive officer (CEO) drily noted that the prosecuting councils were 'trying to reduce our ability to operate' – but the company had a choice: it could take on board the councils' concerns about the environment or it could ignore them. In the end, it decided it would address its poor environmental performance and it embarked on a comprehensive environmental training programme, from boardroom to bulldozer blade.

By 2008, a mere four years later, having done nothing other than implement a thorough environmental training programme, the firm's turnover was NZ\$300 million, and it had 150 projects on the books, three environmental infringements and 2 per cent non-compliance on its site environmental controls. This turnaround was reflected in much-improved non-price tender attributes, credibility and levels of trust with clients. A few years later, the company won an environmental award for a very tricky piece of streamworks. Now, more than ten years after that, the firm continues to win interesting and environmentally challenging projects from big clients based on its excellent environmental track record.

As the CEO now observes, good environmental management makes good business sense.[16]

E-learning for a water supply, wastewater and stormwater utility

Some years ago, a major New Zealand water utility found it was taking far too much time for staff to review environmental management plans for major public works because they were prepared in different formats, with different contents and to widely varying levels of quality. The organization decided to require its contractors to prepare these plans to a new and consistent standard. To be eligible to bid for work with the utility, every contractor had to prepare a company environmental management plan every year. Some large or very environmentally risky projects also needed a specific project environmental management plan. But the plans had to work for the many small contractors, some of them 'one-man bands' such as concrete cutters who carry out their work from a van, as well as for the big consulting and contracting firms.

The utility provided a comprehensive support package for its contractors in the form of an interactive toolbox comprising:

- a leaflet summarizing the package;

- a background document explaining the environmental management plan framework and the reasons for setting it up;

- an electronic plan template with all the headings set out and helpful tips provided for the content required;

- printed workbooks containing background information and examples to help contractors to fill out the template for their organization;

- a set of environmental control procedures and other resources for the contractors to use and adapt for their own site management and monitoring;

- a self-paced interactive online learning programme, supported by classroom-based training and free access to computers in the utility's offices; and

- ongoing support from staff in the form of toolbox talks, industry presentations, a 'helpline' email address and free staff advice when required.

The first plans were uniformly excellent – a result that project sponsor Michael Lindgreen said could not have been achieved cost-effectively other than by training every contractor to prepare their management plans to the same high standard. Regular worksite inspections made sure the contractors were supported in following and, if necessary, reviewing their plans. The number of environmental incidents significantly dropped, while the intangibles of improved industry reputation, contractor buy-in, client–contractor relationships and demonstrable industry best practice increased dramatically.

Erosion and sediment control in the City of Charlotte, North Carolina

Jay Wilson, a Certified Professional in Erosion and Sediment Control (CPESC) who works for the City of Charlotte, NC, described the trigger for Charlotte's erosion and sediment control programme as two 'giant' developments of over 3,000 acres (over 1,200 hectares) proposed near a river with multiple uses, including drinking water supply, recreation and power generation.[17] The threat posed to these waters by sediment runoff was a focus of legal environmental permits, so when the land was re-zoned to allow the development, the permits required the contractors to be trained in erosion and sediment control and general water quality protection measures, such as dealing with paint and concrete.

The steps in the training process were to:

1. assess the need and identify the benefit;

2. identify the target audiences;

3. develop the training content;

4. deliver the training;

5. set up record-keeping systems; and

6. create synergies within and beyond the council.

The audiences that Jay identified, each with their different information needs, included:

- venture capitalists and developers;

- the design community (engineers, architects, landscape and related professionals);

- the construction community (heavy construction contractors, subcontractors and building specialists, including framers, painters, electricians, plumbers, drain-layers);

- city and county staff and third-party inspectors; and

- community and volunteer groups.

Core training materials and resources included local and state standards, guidelines and manuals, plus federal directives. Jay emphasizes that key people can be great resources too, and he asked some of the City's local programme directors, engineers and inspectors to present parts of the training. A useful asset was a photo gallery of the whole gamut of environmental performance, captioned 'The good, the bad and the ugly'.

Delivering the training includes not only running the actual class, but also all the necessary (and time-consuming) organization, as well as the administration that keeps the training programme going, which includes keeping and updating good records.

There are challenges in maintaining the programme, but the rewards and benefits are also there – and statistics on outcomes such as those seen in the City of Charlotte can help you to build your business case:

- an elevated awareness of water quality regulations and why they're needed;

- many of those attending the classes later call in to report water quality problems, meaning that each attendee becomes an extra pair of eyes in the field;

- fewer notices of violation being issued;

- fewer environmental penalties being imposed; and

- better water quality – the overarching goal that the training was set up to achieve.

Voluntary community riparian restoration and enhancement programmes

In 2001, the then Auckland Regional Council released a riparian management strategy that promoted voluntary measures, including community-based initiatives and support to encourage riparian zone improvements.[18] Targeting priority land uses, the strategy's goal was to:

- retain existing riparian zones in good condition;

- enhance existing riparian zones in poor condition; and

- restore riparian zones where they did not currently exist.

The Council offered training workshops to its own staff and to landowners, plant nursery owners and community groups. Trainees were given a free copy of a detailed planting guide, as well as a practical workbook, and were encouraged to bring along maps and photos of their own properties for discussion. They visited a stream needing riparian restoration, where they drew a streambank profile, referred to the planting guide to select plants suitable for the different riparian zones, and drew up a planting plan and map to present to the other attendees for helpful feedback. The great thing about this project was that it built capability in two important areas: the staff who were delivering the training workshops and the people who attended them.

The Council decided that its own staff could best deliver this workshop, and set up a pool of people with expertise in terrestrial and aquatic areas who then attended an 'environmental train the trainer' workshop, which I ran for them. The workshops generated magnificent public relations (PR) for the Council: the trainees asked questions about other aspects of the Council's work and the answers gave people a much better understanding of the many wonderful things it does for the region. I heard many of the trainees saying, 'I never knew the Council did this kind of work – now I can see what I'm paying my taxes for!'

My involvement with this project came to an end with a series of 'train the trainer' workshops to support council staff delivering the training, but nowadays I would insist on supporting the client to develop an evaluation plan that went further than tracking outputs (the workshops) to include monitoring of outcomes – that is, the short-, medium- and long-term increases in planting, weeding and maintenance of plantings and their associated soil conservation, terrestrial and aquatic ecology, and carbon sequestration benefits.

The Digger School: a polytechnic–government partnership

Although this example is no longer current, it is an example of a straightforward path towards a qualifications-based system. Environment Canterbury (ECan), is the environmental regulator for a region in New Zealand's South Island. Its 'education for sustainability' team coordinated an environmental training module in association with the local polytechnic's Digger School. Before they started, the polytechnic and ECan staff had to consider:

- the commitment from ECan and the Digger School;

- the available time of relevant staff;

- job descriptions and work plans;

- the available budget;

- the suitability of staff members and tutors to organize and present the training; and

- the ability and willingness of both parties to work collaboratively.

The training involved a mix of contextual, theoretical and practical training to ensure that every digger operator knew what to do to prevent sediment runoff into the environment. Both ECan and Digger School staff delivered the training in class and on working sites, discussing environmental issues, problems encountered, and how the site was controlling erosion and sediment runoff. Students prepared an environmental portfolio to be judged by ECan and Digger School staff at the end of the course, and not only did they receive recognized tertiary qualification, but also they were eligible to win an environmental awareness award.

Community capability-building: an indigenous peoples example

Community development has always been something of a mystery to me – but, in recent years, I've understood it better, having seen a training course adapted by a Māori tribe and a community engagement programme focused on waterways, both initiatives delivering amazing capacity into their communities.

Several years ago, I helped with a series of nationwide riparian workshops funded by the Ministry for the Environment. They were picked up by Ngāi

Tahu, the Māori tribe that holds the rangatiratanga (tribal authority) to over 80 per cent of New Zealand's South Island – Te Waipounamu (the Greenstone Isle). Craig Pauling, a staff member of the tribal authority, sought permission and resources from the Ministry to run two workshops to make the materials applicable and available to the tribe's local committees.[19]

The shared vision was of weaving a cloak to cover the nakedness of Mother Earth, stripped of her mantle of vegetation. Craig's goal was to transfer restorative skills to rūnanga representatives, including:

- understanding the values and functions of riparian zones;

- planning and implementing riparian restoration and management; and

- training others in these skills – a way of disseminating knowledge to the committees.

I loved Craig's response to someone bemoaning the size of the restoration job ahead of them: 'Well, it took 150 years to remove the trees. So what if it takes us another 150 to restore them? At least we're making a start!'

And the more I think about Craig's comment, the more I like it: that's five or more generations of people learning how to care for the land, increase rural productivity, landscape values and freshwater fisheries – and set up new businesses at the same time.

After the workshops, Ngāi Tahu won the Professional/Institutional Category in the Environment Canterbury Resource Management Awards 2004 for the guidelines and associated work-around riparian management and training.

Similar initiatives with indigenous peoples are being supported in other countries.

Identifying other examples

What if you can't find an environmental training case study in your sector of interest? Try searching online for one from another sector, using search terms such as:

- green [insert business sector], [insert country];

- sustainable [insert business sector], [insert country];

- eco [insert business sector], [insert country]; and/or

- environmentally friendly [insert business sector], [insert country].

Still no luck? Try leaving out the [country] term and search for things such as green fashion, sustainable cafe, eco beautician, environmentally friendly carwash and so on. If you find a sustainability guide relevant to your search, then you can follow it, together with this book, to develop your own training programme – and then be sure to publicize your work, so that others can find it.

NOTES

1 Lessonly.com (n.d.) Definition of training. Available at www.lessonly.com/definition-of-training/ [accessed 15 August 2019].

2 Jessica Miller-Merrell (2012) 41 per cent of companies with poor training have employees leave in 12 months: employee retention vs. recruiting programs. *Workology*, 12 December. Available at https://workology.com/employee-recruiting-retention-focused/ [accessed 15 August 2019].

3 H.S. Timperley, A. Wilson, H. Barrar and I. Fung (2007) *Teacher professional learning and development. Best Evidence Synthesis Iteration (BES).* A report prepared by the authors (all from the University of Auckland) in December 2007 as one of a series of best evidence synthesis iterations (BESs) commissioned by the Ministry of Education. Available at www.educationcounts.govt.nz/publications/series/2515/15341 [accessed 5 June 2019].

4 Jost Reischmann (1986) Learning 'en passant': the forgotten dimension. Unpublished paper. Available at https://works.bepress.com/jost_reischmann/ [accessed 16 September 2019]. You can find out more about Professor Reischmann's views on andragogy at www.andragogy.net/ and www.reischmannfam.de/ [both accessed 16 September 2019].

5 Dr Doug McKenzie-Mohr (2011) *Fostering sustainable behavior: An introduction to community-based social marketing*, 3rd edn. New Society, p. 152.

6 You can find out more about UNITAR's work and training opportunities at www.unitar.org/about/unitar/institute and www.unitar.org/sustainable-development-goals/accelerating-sdg-implementation [both accessed 5 June 2019].

7 Shirley Walters (2019) *Shirley Walters: The power of lifelong learning.* An interview broadcast on Radio New Zealand on 11 June. Available at www.rnz.co.nz/national/programmes/ninetonoon/audio/2018699052/shirley-walters-the-power-of-lifelong-learning [accessed 11 June 2019]. Dr Walters is Emerita Professor of Adult and Continuing Education, University of Western Cape, South Africa.

8 D. Wentworth (2015) *Compliance: not just a training challenge.* A webinar by David Wentworth, senior learning analyst at Brandon Hall Group, and Steve Young, general manager for Asia Pacific at NetDimensions, delivered on 17 November.

9 V. Kenrick (2011) New sustainability professionals within the retail industry. *Environmental Leader*, 9 September. Formerly available at www.environmentalleader.com/2011/09/19/new-sustainability-professionals-within-the-retail-industry/ [accessed 22 November 2011].

10 Bill Cushard (n.d.) *Subject-matter experts and keeping up with the demand for learning.* Available at www.mindflash.com/blog/subject-matter-experts-and-keeping-up-with-the-demand-for-learning [accessed 15 August 2019].

11 'Train the trainer' training is offered by many commercial and not-for-profit training providers and hopefully you will be able to find a provider near you. I warmly recommend this training. I offer a popular 'learning to train' workshop especially for environmental experts and you can find out more about this and other workshops at www.ESST.institute/traininghub

12 J.C.R. Licklider (1965) *Libraries of the future part 1: Man's interaction with recorded knowledge*. MIT Press, p. 17. A predecessor of Bill Gates' quote appears, and was subsequently adapted and popularized by Bill Gates in the Afterword to Bill Gates (1996) *The Road Ahead*. Penguin Books, New York. Based on data from the Yale Book of Quotations and ABC News, according to Quote Investigator at https://quoteinvestigator.com/2019/01/03/estimate/ [accessed 15 August 2019].

13 Gina Abudi (2017) What is your training strategy? Available at www.ginaabudi.com/what-is-your-training-strategy/ [accessed 19 April 2019].

14 The full translation is: 'My success should not be bestowed onto me alone, as it was not individual success but success of a collective.' Said humbly when acknowledged. See www.maori.cl/Proverbs.htm [accessed 15 August 2019].

15 This quote is usually attributed to US humourist Will Rogers (1879–1935), but while there is good evidence that he never (and would never have) said it – see Ben Yagoda (2017) Will Rogers said that. Except he didn't. *The Chronicle of Higher Education*, 12 February. Available at www.chronicle.com/blogs/linguafranca/2017/02/12/will-rogers-said-that-except-he-didnt/ [accessed 15 August 2019] – no other source has been found. See also https://en.wikipedia.org/wiki/Electric_fence [accessed 15 August 2019].

16 D. Adams (2008) Environmental management: the business case. A presentation to the Auckland Regional Council Sediment and Stormwater Field Day on 30 September.

17 J. Wilson (2011) Developing a local training programme. A paper presented at EC-11, the 2011 conference of the International Erosion Control Association (IECA) in Orlando, FL, in February. Find out more about the City of Charlotte's environmental activities at www.charlottenc.gov/StormWater/Pages/default.aspx [accessed 5 June 2019].

18 Auckland Regional Council (2001) *Riparian Zone Management Strategy for the Auckland Region*. Auckland Regional Council Technical Publication 148, June. Available at www.aucklandcity.govt.nz/council/documents/technicalpublications/TP148%20Riparian%20zone%20management%20strategy%20guideline%20planting%20guide%20Strategy%20-%20%202001.pdf [accessed 17 September 2019].

19 Craig Pauling (ed.) (2003) *Riparian planting and management guidelines for Tangata Whenua*. Ngāi Tahu, New Zealand Landcare Trust, Ministry for the Environment, Manaaki Whenua Landcare Research, National Institute for Water and Atmospheric Research (NIWA), Environment and Business Group and Waiora Soil Conservation. Find out more about Ngāi Tahu at www.ngaitahu.iwi.nz/.

3

The Success Framework: the seven elements of successful environmental training programmes

A carelessly-designed project takes three times longer to complete than expected; a carefully-planned project takes only twice as long.

Golub's Second Law[1]

Effective environmental management programmes comprise several essential elements that support each other. Training is only one of them.

The case studies in the previous chapter share many common elements. Figure 3.1 illustrates the seven elements of success I distilled from observing successful programmes. Details of the content may vary, but the elements themselves are consistent across business, government and not-for-profit bodies.

In this chapter, we'll cover six of the seven elements, with the next chapter devoted to the training itself.

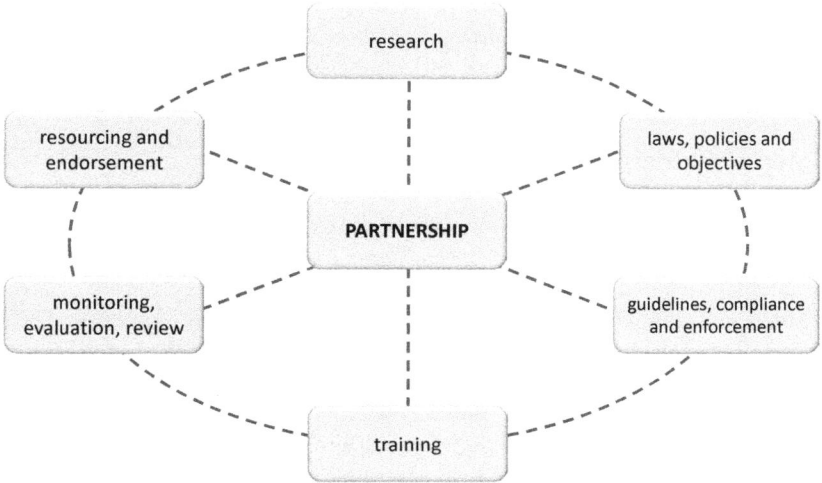

Figure 3.1 The Success Framework: the seven essential elements of successful environmental training programmes

The elements of the Success Framework aren't fully sequential: some can be developed in parallel, while keeping them all mutually reinforcing – but the most important one at the outset and right though programme development, implementation and monitoring is partnership.

 ACTION PLANNER

As you go through this chapter, jot down your thoughts about the elements of the Success Framework on Action Sheet 3.1, which you'll find in the free Action Planner that accompanies this book. The seven elements will broadly apply to businesses and not-for-profits, as well as to government bodies.

3.1   PARTNERSHIP: THE FUNDAMENTAL PLATFORM

> If you want to go fast, go alone. If you want to go far, go together.
>
> African proverb[2]

Partnership is the platform par excellence for an effective environmental training programme, regardless of how strong or weak the regulatory and community focus on the issue in question. We can be so focused on our core technical competencies that we overlook the so-called soft skills that maximize their effectiveness. Partnership is one of these.

Your partners need to be the people, groups or organizations whose support or lack of it may significantly influence your programme's success. Your first partners will be internal, and your programme will involve ongoing work with both internal and external partners and other stakeholders over time. This work is often called 'stakeholder analysis' and, although the term seems rather mechanistic, it's a very important step.

Internal partners

Look for your first partners within your organization; you will find more as you go. Obvious partners include people who are already involved or interested in training, or who are responsible for your organization's strategy, research, monitoring and community engagement, or those who work with environmental permitting and legal compliance. Two great partners often overlooked by would-be environmental trainers are your 'money people' and your 'people people'.

Who holds the purse strings? Your accountant can be a big buddy or a big barrier when it comes to sustainability: they will already have many of the numbers you need to work with and can help you to measure additional finance-related indicators of your training outcomes (see Chapter 4). Make your accounting team your partners by showing them how your environmental training can:

- cut costs, including by reducing staff turnover;

- bring in revenue from environmentally minded customers; and

- increase company value.

For governments and not-for-profits, the same applies in broader terms – especially demonstrating how your proposed training could cut the costs of impacts on the environment and communities, and the costs of continually reacting to these rather than preventing or reducing them.

The learning and development (L&D) staff in your human resources (HR) or personnel team also have a big role to play. Depending on the size of your organization, they will already be delivering in-house training and using a learning management system (LMS) to track who is trained on what and when. Without an LMS, it can be time-consuming to work out what environmental training has been done in the past, who has been trained on what and by whom, what resources are available, who the key people have been, and who is available and suitable to bring on board to help you with your training programme (more on this in Chapter 7). Being able to track your training in an LMS will be an enormous advantage, and you may also be able to enlist your HR team's expert support in developing and delivering your training.

Beryl Oldham, an expert in return on investment (ROI), points to a growing demand for HR staff to support initiatives that closely deliver on organizational objectives.[3] Their input will help you to develop your business case by outlining, in business terms, how your proposed training programme will benefit the organization.

Case study: what constitutes partnership?

Grant Crossett specializes in designing management systems for invasive pest animals and is a pest and predator control monitoring specialist.[4] When I met him at an international conference on education and training, he told me how he works closely with a wide range of stakeholders when preparing and delivering his programmes, including councils, government departments, local community groups and conservation volunteers. He was struck by my presentation because he'd never thought of how he worked as being 'partnership' – but it undoubtedly is. For Grant, it's just been how he naturally goes about his work. Now, however, having an awareness that he is developing partnerships rather than only building work processes will enable Grant to consciously strengthen his working relationships.

External partners

Be fully inclusive: bring in your naysayers, because you need to understand their perspectives and concerns. Rich Batiuk, recently retired after 33 years at the massively successful Chesapeake Bay Program, said that creating an inclusive management structure was one of six core lessons he learned over three decades (more on these in the last part of this chapter). 'You need to build a big enough tent' to hold all parties, he said, and when asked if there was one thing he would have done differently earlier knowing what he knew now, he replied, 'I'd have built a bigger tent to include more partners from the start.'⁵

That said, for some contested issues, you may just have to start with only a few willing participants to get some initial traction and keep all walls of the tent open for others to join you over time.

External partners and stakeholders are very specific to the issue and sector you're working with. Industry, professional and trades associations, including unions, are extremely helpful and often have training programmes under way that you can support or supplement.

Businesses and not-for-profits bodies may work with external partners in developing and delivering environmental training. This is commonly through external training providers, including universities and technical colleges, with the local environmental agencies requiring or encouraging this training, or with local community interests. For government bodies, I have found partnership to be crucial to the success of compliance-based initiatives such as environmental training.

Approach the 'just transition unit' for your country or state and work along-side them. If you don't have such a unit yet, consider working with the sector's union and the government agency responsible for supporting people into employment. I've seen great interest from such agencies in environmental training for entry-level employees, because environmental training creates green jobs.

It's never too soon to start. If you need to gather environmental research to define the environmental impacts of your sector of interest, then involve the sector in defining the research questions and methods; if you're developing policy and regulation, involve them in that too. If you're creating or adapting a technical guideline, get the sector's input and feedback. When you're developing your training, ask them to help you to pilot it before rolling it out. Yes, it may take longer up-front, but – based on my observations of several similar initiatives in which environmental regulation and possible legal enforcement were involved – early engagement and partnership will save years of battle further down the track.

A good way to do this is to set up a liaison, advisory or working group to represent the major stakeholder groups and provide a way of communicating with the wider sector. For example, with Auckland's erosion and sediment control training programme, the liaison group included staff from the regional council (equivalent to a state-level environmental protection agency), a city council, a developer, an engineering consultant, a civil engineering contractor, a tribal representative and an official from a large community environmental group. Rapid urban growth at the time meant there were tensions between all the parties, but they came together with goodwill (and even better catering) and, over time, became a highly cooperative and effective group.

You could draw up an informal brief that may evolve into more formal terms of reference, but keeping at least the first few meetings informal will help your programme to become genuinely collaborative. As trust builds over time, this kind of forum also makes it easier to broach sensitive issues not easily resolved by more formal avenues.

Developing a constructive and collaborative partnership between regulator and industry ensures that both regulator and industry can meet their own needs while remaining aware of each other's, and hence may be able to avoid framing regulation that is unduly prescriptive and burdensome to both parties.[6] This requires formal and informal mechanisms for ongoing communication – communication that involves listening as well as talking. Long-term commitment is also essential to avoid cynicism and foster genuine commitment to improving process improvements by the regulators and environmental performance by the sector.

It's helpful to identify your engagement with and leverage over the sector as part of scoping partnership opportunities, for example by asking the following questions.

- How well do you know and get on with its leaders and members?

- What trust or other issues are there among you?

- How aligned or polarized are views on the issue?

- How will they be likely to respond to your idea?

- What will they want to know about how it will affect them?

- How can you make a good case for their attending your training? For example, you might:

  - make such attendance a requirement for tendering for any work;

  - refer to the training guidelines as a performance requirement for obtaining environmental permits; and/or

  - build your legal ability and staff capacity and capability to use enforcement or financial penalties or incentives to encourage compliance.

Partnerships take many forms. Sector associations working with government bodies in long-term partnerships to deliver recognized environment training are common. Some may progress over time to a formal memorandum of understanding (MOU) and I've seen environmental MOUs signed between:

- a council and an indigenous tribe;

- a council and a training provider;

- a business and an environmental not-for-profit body;

- a council and an environmental not-for-profit; and

- a not-for-profit, a council and a training provider.

What if your sector is disorganized or polarized?

You don't have to wait until everything is perfect before you can start. One colleague told me of an example in which there was a real environmental issue, but the industry sector of concern was very poorly coordinated. So that it could have a representative body to work with, the state environment regulator contributed seed funding and staff support for two or three years until the sector was fully functional and independent, and able to enter into an effective dialogue with the regulator to tackle the problem.

Earl Shaver recounts another great example about the erosion and sediment control programme in the state of Delaware in the United States.[7] It lacked resources and operated in a context of inadequate legislative and regulatory authority, and there was no strong environmental lobby group to represent the community's growing concerns about flooding and erosion. Earl and his colleagues at the Delaware Department of Natural Resources and Environmental Control developed a consensus-style approach to getting the necessary legislation and subsequent regulations accepted by the state legislative bodies and by the industry that was going to be regulated. They developed an education campaign highlighting the size of the problem, with slides showing both the impacts and the nature of the regulatory and training solutions. They rolled out the slides in meetings with and presentations to contractors' associations, engineering consultants, utility companies, land developers and the general public. They also made informal presentations to legislative committee members and received only one negative vote in the two-stage process of voting in the new legislation. This approach was so effective that not one member of the sectors affected submitted or testified against the regulation that would affect it. The continued success of Delaware's programme is a tribute to the consensus approach.

Sometimes, however, if the stakeholders have become very polarized, preparatory work is needed before the parties can constructively engage with each other. The big environmental issues facing us today often generate deeply opposing views about how to address them, even when all parties agree on the importance of the problem. In everyday life, we normally like to work with people we know, like and trust – but, as Dr Enette Pauzé says, today's environmental issues are so complex that we often need to work with people we don't know, don't like and don't even necessarily trust.[8] If this is the situation in which you find yourself, enlist the best help you can find to build the lasting partnerships you'll need.

Community partners

Community development has often emerged as an unexpected result of place-based environmental programmes, but it's increasingly being planned for as an outcome in itself by replicating successful locally inspired initiatives such as community gardens in economically deprived areas.[9] Many of them have succeeded in encouraging local people to gain trades and professional qualifications to which they would never have otherwise aspired. Such local initiatives have grown from the ground up for reasons as diverse as a doctor's prescribed anti-dote to patients' clinical depression, the beautification of bleak urban areas or as alternatives to gangs for locally unemployed youth.

In other cases, councils can take a community development approach as one of a range of methods of achieving environmental outcomes. They can choose to work with and provide financial support, expert advice and other support for community organizations and local people to achieve programme goals by working with the council to organize community events, deliver public education training, and carry out practical works such as riparian planting and control of plant and animal pests. In one case, many local coordinators became so skilled and competent that they gained full-time employment with the council and other organizations.[10]

Whether initiatives start from the grass roots up or bureaucracy down, it is important to take a long-term view of how best to support, scale up and/or replicate such work without rendering it vulnerable to shifts in funding priorities of any public funding and support.

 ACTION PLANNER

Use Action Sheet 3.2 in the free Action Planner that accompanies this book to list potential internal and external partners who will want or need to be involved in your programme and carry out a stakeholder analysis. Be as inclusive as possible.

3.2 RESEARCH: BUILDING A ROBUST CASE FOR THE ENVIRONMENT

> They say a little knowledge is a dangerous thing, but it is not one half as bad as a lot of ignorance.
>
> Granny Weatherwax[11]

Good research is the foundation stone of cost-effective environmental management programmes, including their training component.

Routine environmental monitoring can signal trends or changes that justify further targeted investigation and intervention; research can then be done into the effectiveness of actual or potential management methods and their effect on the environmental issue of concern.

Research helps you to define the issue, identify the environmental outcomes you want, identify the most cost-effective methods of achieving them and frame all these in a way that enables their monitoring. It includes initial and ongoing research, and provides data to inform your programme monitoring and evaluation. Involving your partners will help you to make sure that the research results are relevant and credible to them.

How to define an environmental issue

Environmental issues may be existing or potential and can include the cumulative effects of activities over time. They may also reflect the absence or failure of policy. How issues are defined needs to be based on evidence, including evidence gathered by consultation.

The more clearly you can define the issue – preferably by consensus with your internal and external partners – the easier it will be to measure and manage it. An issue statement should provide a baseline measure of the state of the environment – the point you're starting from and against which you will measure improvement. The issue statement should be succinct, with supporting explanations set out in subheadings or appendices, and should clearly identify:

- what the issue is – what is being affected and how it is being affected, what the opportunities are for improvement;

- why it is an issue – what circumstances give or gave rise to the issue, what the cause of the issue is or the scope of the opportunity;

- for whom it is an issue – who the directly affected and the more broadly interested parties are;

- the spatial scale of the issue – where the issue is, what its existing and projected geographical extent is;

- the temporal scale of the issue – when it does or did occur, whether it is intermittent in nature or relates to a specific time frame or event, whether it is ongoing or not, what its duration and frequency are; and

- how urgent the issue is.[12]

Robust research can take a year or more and should address issues and management options in other places – including their efficiency and effectiveness – to help you to build a case strong enough to justify setting up a training programme.

You could also use PEST and SWOT analyses (see Chapter 2) to regularly scan for potential or emerging issues that training could help to address.

Baselines and benchmarks: from where to how far?

Baseline data answers the question 'Where are we now?', while benchmarks help to answer the question, 'Where do we want to be?' Both are essential for monitoring progress. Together, they also inform the setting of objectives that are not pointlessly low or unrealistically high.

It seems blindingly obvious that we need baseline data before we start our interventions – but this practice is not as common as it ought to be! In my observation, initiatives that start from good baseline data tend to be in sectors that routinely measure key indicators, for example air or water quality, the solid waste sector or water and energy utilities. It seems to be more difficult for programmes that involve natural processes, such as erosion or revegetation, to do this – but we need to make the effort or we won't be able to work out very accurately what difference our programme makes.

Wherever possible, we need to define our outcomes in terms of measurable indicators rather than perceptions, which often reflect assumptions or prejudices we may not even know we have.

Remember too that geographic variations may indicate the need for different benchmarks when setting, for example, low-flow levels/durations/frequencies, water quality and other ecological outcomes.

Benchmarking can:

- tell us how much improvement we need to make, derived from research, best practice or guidelines; and

- help us to compare our programme with one or more others – which will encourage us all to do better.

Why not benchmark your programme with an equivalent programme from somewhere else that's in the vanguard of best practice or one that's at the learning stage, like your own?

 ACTION PLANNER

Use Action Sheet 3.3 in the free Action Planner that accompanies this book to define your environmental issue of concern as tightly as you can. This will help you to define the baseline status that your environmental training programme aims to lift. Use the same sheet to jot down ideas about what benchmark you want to or must reach.

3.3 LAWS, POLICIES AND OBJECTIVES: THE GOALS TO ACHIEVE

> We need clear definitions and goals – if you aim at nothing you will hit it every time!
>
> Zig Ziglar[13]

Policies, objectives and compliance inform both business and government activities. Government bodies also have recourse to significant compliance powers under their environmental legislation.

The work you'll do in Chapter 4 to demonstrate how your training aligns with your organization's espoused goals will further develop this element of the Success Framework by clarifying and prioritizing the policy outcomes your training must help to achieve. Here, I focus on:

- meeting commitments to espoused goals, such as international treaties for climate change, as well as national and local regulatory and policy requirements;

- meeting your organization's measurable objectives and targets; and

- compliance, in the sense of:

  - bringing your staff up to a consistent level of performance; and

  - bringing external parties up to the desired performance level, such as firms along a business supply/value chain or companies with governmental compliance oversight.

We'll look at enforcement later in this chapter.

How to demonstrate alignment with espoused goals

It's not easy to clear a path through the thickets of legal and strategic environmental requirements. But it makes a strong policy case when you do.

Here's the thing: your training will have beneficial effects that go far beyond the environmental outcomes that are your core concern. Taking an informed approach to your policy context will help you to discover these – and measure them.

Figure 3.2 shows the many layers of documents that require and guide governments' environmental actions, from global to local. Many of them provide detailed sets of indicators against which to measure progress.

INTERNATIONAL
- UN Sustainable Development Goals
- UN Framework on Climate Change
- other treaties, conventions, protocols

NATIONAL
- environmental legislation
- research, policies, guidelines
- measures of the four well-beings

REGIONAL
- policies and guidelines
- plans, bylaws, codes
- growth, biodiversity and other strategies

CULTURAL
- tribal management plans
- partnership/co-governance agreements
- cultural well-being and other monitoring

LOCAL
- district plans, rules and bylaws
- asset management plans
- parks, roading, waterways, other plans

TECHNICAL
- environmental and other research
- watershed management plans
- environmental permits and compliance

Figure 3.2 Layers of guiding and requiring documents for government bodies

You need to assess these documents to ensure your environmental training programme aligns with and gives effect to them. It can be a complex task: a colleague and I found 120 documents relevant to the outcomes required from one small urban watershed management plan! However, once done, the analysis paves the way for others and updates are easier.

Expanding on the international commitments illustrated in Figure 3.2, Figure 3.3 shows the United Nations Sustainable Development Goals (SDGs).[14] Each goal is accompanied by a detailed set of indicators and governments worldwide have signed up to use them to measure their progress. Businesses and government agencies are now showing how their activities support progress towards the SDGs. Environmental training programmes deliver beneficial outcomes across many of them and should also demonstrate these links. This is usually done simply by placing the relevant icon(s) and the accompanying summary under the heading of each key outcome of your training.

Figure 3.3 Mapping your training outcomes against national and global indicators. *Source*: UN Sustainable Development Goals Knowledge Platform, https://sustainabledevelopment.un.org/topics/sustainabledevelopmentgoals [accessed 5 June 2019].

 ACTION PLANNER

Does your organization align itself with international, national and local goals? Are your work programmes required to demonstrate how they contribute to particular espoused goals? What frameworks or processes are available to help you to do this? Use Action Sheet 3.4 in the free Action Planner that accompanies this book to work out how you can align your training programme with your organization's required or desired outcomes.

Business and not-for-profits

Aligning your training outcomes with your organization's business improvement objectives, your policy analysis and the SDGs will help you to prepare the business case for your training, which we'll look at in the next chapter. For now, you need to look at the job descriptions and key performance indicators (KPIs) for staff with environment and sustainability responsibilities to ensure that they are required and empowered to carry them out. You will need the support of your manager and your HR team to do this, and it's a very important part of building your Success Framework. You will also need their help to foster the willingness and capability of key staff to carry out their environmental roles.

Supply chain management is a specialized area, if you have identified your suppliers as a big part of your environment and sustainability solutions, but a lot of information is now available about greening supply chains. All I would add is the desirability of taking a partnership approach to working with your stakeholders both up and down your value chain.

Government

After doing the research to demonstrate there is a problem to be managed and that training is part of the solution, government bodies sometimes find they lack the specific policy instruments and/or procedures to require members of the sector of interest to adopt new practices and to encourage them to attend training to help them to do so. Common examples are:

- manufacturing and recycling or waste-handling activities that pose a risk of air, water or soil pollution and operate under land-use provisions that are out of date and/or allow pollution response, but don't readily enable a mandatory sector-wide pollution prevention programme; and

- large-scale land development activities, which can be required to obtain environmental permits and use the appropriate erosion and sediment controls, but which are often followed by builders who have no such environmental requirements, often because they are handled by a different government agency or section of the same agency and under a different piece of legislation.

Addressing these policy and regulatory issues to make sure that all the relevant activities pass through the right level of regulatory assessment can take a lot of time and effort over several years, so working with your partners early on to get clarity on what's needed will make a big difference.

 ACTION PLANNER

Use Action Sheet 3.5 in the free Action Planner that accompanies this book to assess whether your policy and/or regulatory provisions support the introduction of a training programme and enable you to require and/or persuade your target sector to attend your training.

Third-party verification

Anyone can put a picture of a dolphin on their website or product label – but the time for such simple greenwashing is long gone. There are now many accepted systems for independent verification of how companies manage their environmental aspects and impacts, ranging from generic systems such as the ISO 14001 standards for environmental management systems to specialized and often nationally specific certification systems for greenhouse gas emissions, product quality, solid waste reduction, green buildings, sustainable infrastructure and more.

The literature is vast, so here I will say only this:

- as part of your business case or ongoing programme development, find out what accepted environmental management and reporting systems may be relevant to your sector;

- consider the value case or business case within your organization of gaining third-party verification of your environmental claims or those of your sector of interest; and

- consider the many reputational, communication and marketing benefits of being able to validate your environmental claims.

3.4 GUIDELINES AND COMPLIANCE: A PERFORMANCE BENCHMARK

> Culture drives behaviour – not rules or guidelines. We are vulnerable to the social norms of our environment.
>
> David Gebler, Ethics Engagement and Integrated
> Education at Lockheed Martin[15]

Good guidelines and procedures set the benchmark we aspire to. Let's not just come up to the environmental minimum; let's see how close we can get to optimum sustainability.

The term 'training', as indicated in Chapter 2, implies the existence of a measurable performance standard or benchmark. In a context in which environmental performance may be required under legally enforceable permits, by complex technical operations or within detailed reporting standards, a detailed specification of requirements and tasks is essential to make it clear what people must do to comply with them.

Many environmental management systems need both structural controls and non-structural procedures and processes. Both can be set out in a technical guideline with the aim of helping the sector of interest to comply with new policies, regulations and/or performance standards.

Technical guidelines are not usually legal documents in themselves, but they effectively become enforceable if attached to an environmental permit or referred to in a statutory plan. While they can and do change over time to reflect new knowledge, these guidelines define the current best practice against which training and both structural and non-structural measures can be audited. This in turn enables evaluation of the effectiveness of the control measures and also of your training.

As environmental experts, we're understandably keen to leap straight into the technical specifications of our guideline, be it for streambank planting, water pollution control, erosion and sediment control or whatever our topic of concern may be. But it's a great idea to step back and think about our pet guideline from a wider perspective, such as by asking ourselves the following questions.

- Who's going to use it?

- Why would they believe us when we say they should use it and that it will work?

Development of a new guideline from scratch is a big project, even when there are many good examples to follow. Follow best partnership practice: involve your sector partners, experts from your research and tertiary institutions, and other stakeholders right from the start. They help us to weed out those unrealistic assumptions, bureaucratic and confusing language, and errors that have faded into invisibility as we become overly familiar with the working drafts.

My work in preparing guidelines for scrap metal yards, septic tanks, panel-beaters, industrial pollution prevention, wastewater overflows, and erosion and sediment control has repeatedly demonstrated to me the significant benefits that partnering yields in both the technical quality and industry acceptance of the new requirements.

Recently, my colleague Graeme Ridley asked me to think about what 'contemporary international best practice' for an environmental control might mean.[16] It was easy to find a good relevant definition for this. The US Environmental Protection Agency (EPA) states that best practice consists of 'scientifically sound techniques [that] are the best practices known today'.[17]

To this, Graeme adds that every guideline needs to be read and followed with knowledge and experience that rounds out the bare technicalities in the geographic context of each worksite. Blind adherence can lead people to overlook important aspects of local context and here, again, good communication between regulator and regulated is crucial.

Criteria for a 'good' guideline

Apart from the best practices that we want it to contain, what else is important for a guideline? What are the criteria for a 'good' guideline? Much of the thinking in this field is medical, but I believe it applies equally to environmental guidelines.

Table 3.1 sets out 14 criteria for a good guideline that I've adapted from two medical sources.

Table 3.1 14 criteria for a good environmental best practice guideline

Criterion	Definition
Inclusive	Representative and multidisciplinary input from key stakeholders in a genuinely consultative process is used to develop the guideline. This is in line with the partnership principle at the heart of good environmental management and training.
Evidence-based	Valid research and the best evidence is used to justify the guideline and its contents. Details of the evidence base are available, and the cause–effect and cost–benefit relationships and assumptions are clearly set out to show how implementing the guideline can be directly linked to avoidance or minimization of the adverse environmental effects of concern.
Relevant	The contents of the guideline reflect local and national issues, laws and policies that are relevant to the guideline's intended users, and changes to laws and policies are accommodated as needed.
Reproducible	Other groups would come to the same conclusion after examining the same evidence.
Robust	Draft versions of the guideline are reviewed by scientific and technical peers and lay users.
Transparent	If users are in doubt, they know where to get more information, including the data, assumptions and analysis used to develop the guideline, and how to contest these if necessary. For example, where controls relate to local environmental conditions, data, calculations and assumptions are clearly set out in the guideline so as to be adaptable and contestable.
Independent	Areas of debate and conflicts of interest among stakeholders, and their outcomes, are recorded and kept accessible for reference.
Practical	A good guideline helps users to solve their real-world problems and makes it easy for them to select the right structural and/or non-structural measures to address them.
Understandable	The guideline uses clear and unambiguous language, photographs with interpretation of what is being shown and why, and simple, accurate diagrams that all intended users can readily understand – remembering that some users may have language, literacy and/or numeracy difficulties.
Practicable	Writers and users of the guideline and other stakeholders all understand the barriers to its use and the resourcing implications of its use for both users and inspectors/auditors.
Cost-effective	The guideline helps to reduce the inappropriate use of resources, including time and money.
Accessible	The intended users of the guideline are consulted about the best ways of presenting it and the guideline is well disseminated by a range of methods.
Auditable	The technical content of the guideline can be used to develop clear audit criteria for assessing the performance of its users.
Reviewable	The guideline is regularly reviewed in light of ongoing research and use.

Source: (1) V.T.S. Airedale (2010) *What makes a good clinical guideline?* A PowerPoint presentation in pdf form dated January 2010, no longer available online. (2) Author unknown (n.d.) *What makes a good guideline?* A presentation accessed in May 2012 on the website of Health Protection Scotland, a division of NHS National Services Scotland, but no longer available online.[18]

ACTION PLANNER

Use Action Sheet 3.6 in the free Action Planner that accompanies this book to benchmark your existing guideline against best practice or to make sure you have the best people and processes in place when you create or update a new guideline.

Numbers that count: a simple scoring system for auditing on-site environmental controls

Auditing environmental performance has become a specialized area often associated with high-stakes compliance with international standards and similar requirements. However, an important aspect of auditing is effective communication of the results to stakeholders in terms that busy people can readily grasp. This is particularly important on horizontal and vertical construction sites and in other fast-moving activities.

This section doesn't apply to the rigorous auditing done against ISO standards or other third-party auditing, but summarizes a simple scoring system that made a massive contribution to the success of Auckland's erosion and sediment control training programme. The council developed a scoring system as part of a standardized approach to the weekly on-site inspections of erosion and sediment controls, and it aimed to support consistency across the compliance monitoring results from different site inspectors working in different areas. It awards each of the design and placement, construction, operation and maintenance of each environmental control a score on a four-point scale (1 = great; 4 = verging on the prosecutable):

- **category 1** means that the control is fully compliant with the guideline and any variations required by an environmental permit;

- **category 2** means that there is evidence of very minor actual or potential environmental impact, so that some remedial action is needed to comply with the guideline;

- **category 3** means that there is evidence of an actual, or potential for a, major environmental impact, or that there is repeat non-compliance of category 2 and remedial action is needed; and

- **category 4** means that there is evidence that a major environmental impact has occurred or that there is repeat non-compliance of category 3.

Table 3.2 illustrates how the audit criteria for each category are set out.

Table 3.2 Scoring criteria for environmental controls

Category	Design and placement	Construction	Operation	Maintenance
1				
2				
3				
4				
Totals				
Weightings				
Overall site score including weightings if applied:				

Source: After Auckland Regional Council (n.d.) Miscellaneous documentation.

The system proved so effective that it has been widely adopted in other parts of New Zealand, with local adaptations and improvements, and contracting firms use it to self-screen their compliance ahead of the council inspectors arriving on site. The table could be adapted to include non-structural controls, such as systems and procedures.

The scoring system yielded very informative data, showing which erosion and sediment controls were most often used, which were most often problematic and, of these, which aspects were most often the problem. For example, for sediment retention ponds, the most frequent problem was poor construction, followed (in order of decreasing frequency) by poor maintenance, inadequate documentation (for example having no as-built drawings), not being built at all when required or poor reinstatement after decommissioning.

The site scoring system enables councils and companies to develop KPIs, collate the results into a report, determine overall company performance, and then focus on training and other opportunities to improve poorly performing operators or specific controls as necessary. This model has been invaluable in enabling some very fine-tuned identification of common design, construction or maintenance faults and improvements are then fed into guideline updates.

This information in turn informed not only the content of the training workshops, but also the council's applied research into simpler, more robust and more effective devices. Former programme manager Roger Bannister said it never failed to amaze him how such a simple tool had made such a big difference in site performance.[19] This simplicity made it easier for people to communicate exactly where improvements need to be made to a given control.

Some elements of the scoring system changed over time and in different places. The following are two particular examples.

- Critics said that the early practice of awarding an overall site score based on the lowest score for an individual control was unfair if there were many good controls on a site and the lowest-scoring one was comparatively minor. Based on experience, some weightings have been introduced that reflect the overall quality of controls.

- Some councils carried out an extensive moderation process to make sure that scoring by different inspectors and different councils on either side of an administrative boundary that crosses a site is consistent, because many consultants and contractors work in different parts of the country.

While initially horrified at the introduction of the scoring system, companies now make strenuous efforts to gain and maintain high scores as a measure of their own performance and have carried it round the country to use on their own sites. The scores are widely used for:

- rewarding teams for good performance – some organizations encourage friendly rivalry between teams on different parts of big sites or even on different sites, or award movie tickets or pizza vouchers to the winners, or make a donation to a school or charity nominated by the winning team;

- conveying performance information to the sector, senior managers and elected representatives in a readily understandable format;

- offering project bonus payments as an incentive for good environmental performance and (less frequently) imposing financial penalties for poor performance;

- requiring or demonstrating a good track record on environmental compliance when calling for or responding to tenders for new projects; and

- identifying training needs as they arise.

This system worked very well for the large numbers of devices on big, fast-changing civil construction sites with the potential to cause significant environmental harm if any of the devices were to fail. How well would it work for other sectors, such as solid waste minimization, manufacturing or ecological restoration?

 ACTION PLANNER

Use Action Sheet 3.7 in the free Action Planner that accompanies this book to populate Table 3.2 for the performance of environmental controls for your sector of interest. How could you create an auditable benchmark for defining environmental performance outcomes and scoring workplace performance for that sector?

Changing the compliance conversation: from confrontational to collegial

Too often seen as a cost of doing business, in-house and external audits and compliance monitoring have the potential to be so much more. As we have seen, compliance is an integral part of a training programme's effectiveness. Embraced as a business development tool, it can shape your learning culture and grow your company's expertise and reputation – and its profitability – as a result.

Lawrence P. Leach talks about the difference between compliance and commitment and what it means for project success.[20] In Table 3.3, I've adapted his thinking to show how environmental compliance can be transformed into collegial dialogue and mutual learning among the parties involved in a construction project and their regulator.

Table 3.3 From compliance monitoring to collaborative learning

Compliance can feel:	Dialogue and mutual learning emerge when:
• like being 'told what to do'	• the parties are focused on goals, solutions and outcomes rather than problems
• like being told 'just do it or else'	
• confusing, because the outcome or purpose of the environmental control measures or their inspection is unclear	• all the parties seek to understand both the business and the environmental point of view
• like 'just one more thing to do' that takes time and money, but adds no real value to the project	• compliance is seen as a tool to improve the quality and efficiency of work
• top-down, because there's no chance to discuss, debate or contest audit findings	• compliance is clearly aligned with the organizational goals of both regulator and regulated
• negative, because feedback is not given in a timely way	• compliance is seen as a collaborative effort that is essential to project success
• demanding, because it's 'not part of my real job'	

Source: Adapted from L.P. Leach (2005) *Lean project management: eight principles for success. combining critical chain project management (CCPM) and lean tools to accelerate project results.* Advanced Projects Inc., p. 41.

It's clear that making the transition from compliance to collaboration might require a big change in attitude of some parties, but the benefit is that it reduces legal and environmental risks. Better understanding compliance as a professional development tool can improve working relationships and staff engagement – and can generate the associated increases in efficiency and productivity. And, of course, it doesn't preclude the option of later enforcement action when this is genuinely needed.

 ACTION PLANNER

'Audits should not be considered as a negative exercise … [but] as an opportunity for the continual improvement of the business', says business adviser

Gordon Anderson.[21] Use Action Sheet 3.8 in the free Action Planner that accompanies this book to test your audit processes against Gordon's five straightforward criteria for a constructive audit.

How to achieve consistency across compliance inspections

If you're introducing a system of site inspections to check whether your trainees are following the guideline, make sure your inspectors are as consistently well trained as possible. I have heard concerns expressed over many years by legal commentators and people in many different sectors about the lack of training and experience among council and contracted inspectors. What bothers industry is not the inspections themselves, but getting inconsistent results from the same inspector, different results from different inspectors visiting similar sites or different results from inspectors on similar sites in different parts of the country. In the worst case, this could mean breaches of natural justice, such as taking enforcement action that could have legitimately been avoided. This means that both government and in-house business compliance inspections, vital for promoting good environmental outcomes, lose credibility.

Your environmental training must include your inspectors and your industry trainees. Roger Bannister set up a moderation process for environmental site inspectors contracted to the Auckland Council. He got the idea from the judiciary, where a similar process was put in place to ensure that different judges set similar penalties for similar cases, with the aim of 'getting them to a point of reasonable proximity'. Roger said his moderation process wasn't very sophisticated, but was nonetheless quite effective.[22]

 ACTION PLANNER

Use Action Sheet 3.9 in the free Action Planner that accompanies this book to see how you might set up a simple moderation system for your environmental inspectors.

3.5 ENFORCEMENT: BALANCING FAIR WITH FAST FOR POSITIVE ENVIRONMENTAL CHANGE

> Creating effective compliance training can cost a bit more up front, but it pales in comparison to the cost of doing compliance training poorly.
>
> Sheri Winter, Caveo Learning[23]

Government bodies have recourse to significant compliance powers under their environmental legislation. Perhaps counter-intuitively, when introduced as part of a partnership-based programme, enforcement is often welcomed by the good performers.

Not surprisingly, people feel strongly about enforcement: they either love it or hate it. I believe that enforcement – the use of legal sanctions for breaches of environmental law – is a vital part of the environmental management toolbox. However, it's not a tool of first resort. It needs to be part of a multipronged approach to promote rapid improvements in environmental practice – an approach that must also be transparent and in line with natural fairness and justice.

Understanding regulatory institutions and practices

A fascinating and insightful report into regulatory institutions and practices found that:

> From outside an organisation it can be hard to distinguish problems arising from a 'dysfunctional culture' from those arising from more tangible factors such as the design of legislation or capability issues. When looking to improve regulator performance, it is important to understand if what is required is a change of practice within a given culture, or a change in culture.[24]

That report contains much valuable information and guidance, and emphasizes the importance of accurate definition of the issues and their causes. We'll come to these in Chapter 5.

A useful distinction between regulatory practice and the quality of the regulations themselves is offered by Cary Coglianese in Professor Malcolm Sparrow's 2000 book, *The regulatory craft*.[25] Coglianese grouped regulatory reform issues into the following four major subject areas:

- the **scope of regulation** – including issues such as deregulation, reregulation and regulation of emergent risks;

- the **nature of regulation** – including alternative forms such as the use of tradeable permits in pollution control and negotiated resolution procedures;

- the **locus of regulation** – namely, questions of centralization or decentralization and levels of regional or local governmental autonomy; and

- **practice (behaviour)** – covering the strategies, tactics, policies, operational methods and culture of regulatory agencies.[26]

The key in relation to practice is that 'it falls to regulatory agencies to implement' the three other subject areas and so, as Professor Sparrow observes, the 'style and nature of their implementation can surely make or break any new set of rules'.[27]

Fair and fast – or slow and unfair? Judicious use of enforcement in the Success Framework

Environmental laws usually provide a range of enforcement procedures ranging from the equivalent of a speeding ticket up to prosecutions that attract potentially heavy fines and even imprisonment. Fairness thus becomes a big consideration when it comes to considering the use of legal enforcement as part of an environmental training programme.

For example, it would clearly be a breach of natural justice if an environmental agency were to lack clear standards, but use punitive enforcement in the event of breaches of the law, especially where there were no clear guidelines as to what measures would have enabled the subject of the enforcement to avoid it. Industry would understandably lose confidence in environmental laws and processes. But what if enforcement were not used at all, but good guidelines and industry training were available? In that case, we'd probably see better environmental performance emerge – but it might take a very long time, with valuable environmental resources being damaged along the way and a loss of community confidence in the law.

On balance, then, I think that having a good regulatory framework, accompanied by a good guideline and industry training in how to use it, and the appropriate use of sanctions and incentives offers a faster track to environmental protection. These are, of course, the elements of my Success Framework, as illustrated in Figure 3.1. All the evidence tells me that taking a partnership approach to this issue is highly cost-effective and produces better overall results for participating businesses and the environment in a shorter time.

Does this mean that all prosecutions can be avoided by governments delivering good industry training? Not necessarily. But there's no need to shrink from it when properly introduced. I well remember how astonished we all were at the development industry's generally positive response to the use of legal enforcement, in the form of prosecution, to ensure compliance with the council's erosion and sediment control guideline. Responsible operators felt relieved that the 'fly-by-night' operators, who undercut prices by skipping or skimping on environmental controls, were finally being called to account.

Case study: they really didn't know what they didn't know

As Jay Wilson discovered, when setting up his South Carolina erosion and sediment control training programme (see Chapter 2), although there were serious problems with sediment runoff from developments, the people working on the construction sites thought they were doing what they needed to. They weren't intending to sidestep or break the law, but they didn't understand the principles and processes of erosion and sediment control, and they didn't know what the on-site measures were meant to do or how they really worked. The result was that they were spending time and money in good faith, but building environmental controls that were ineffective.

We need to keep such examples in mind when we begin to use enforcement measures as our training programme gains more traction. An informal 'grace period' can generate much goodwill.

While our early trainees hadn't thought about or didn't like sediment running off sites into streams and onto beaches, they thought it was an inevitable consequence of progress and that it wasn't possible to do anything about it. We got a strong sense of their relief that finally there was something they could do to manage what everyone had thought was an intractable problem. Even when the responsible operators were penalized for some reason, they took it in good faith – and this reflected all parties' commitment to the partnership approach.

So was the use of enforcement an essential element of the programme's success? Yes, it was. And it was all the more effective by being only one element of the seven in the Success Framework. So what would happen if there were a supportive framework, but no prosecution? Or prosecutions without any support for industry to improve its performance? Figure 3.4 outlines some options.

Figure 3.4 Fair and fast – or slow and unfair? Judicious use of enforcement in the Success Framework

The Success Framework sets up a process that is fair to all parties, including communities affected by the environmental issue of concern. Enforcement provisions are ideally supported by a transparent and consistent approach to making decisions about when and how to use enforcement. Some regulatory agencies publish their enforcement decision-making processes and criteria online,[28] which I think should be the norm.

So when is it a good time for a company to invest in good environmental controls? One, when you are heavily regulated and, two, when you're not. Why? Because hard-nosed business experience shows that taking sensible precautions to protect the environment can reduce operating costs and build capability ahead of the introduction of stronger regulation, such as for greenhouse gases, solid and hazardous waste, as well as generally being good – and profitable – business practice.

Table 3.4 offers more detail about the four approaches that government bodies might choose to take to using enforcement, and about how good regulation and enforcement in the context of the supportive Success Framework promotes sustainable changes in practice – more quickly and more cost-effectively.

Table 3.4 Fair and fast – or slow and unfair? Judicious use of enforcement in the Success Framework

Option	Success framework in place?	Enforcement used?
Hanging judge	✗	✓

It would clearly be a breach of natural justice if a regulator had no clear regulatory framework, but used punitive enforcement in the event of, for example, unauthorized discharges to air, land or water – especially where the sector had no clear guidelines as to what measures would have enabled it to avoid them. Some kind of regulatory framework is thus vital to ensure that businesses need some form of legal approval for their operations and have environmental performance requirements spelled out in these approvals. For example, I've seen cases in which regulators are set up to capture applications to discharge stormwater from a completed development, but have no clear legal mechanism to address the issue of sediment runoff from exposed soils during the development process. It can take a long time to introduce or change such requirements; regulators sometimes need to find work-around solutions and should partner with the sector on these. Ideally, the regulator would also have or recommend a technical guideline to help firms to step up to new requirements of plans and permits.

Option	Success framework in place?	Enforcement used?
Wild West	✗	✗

This is the 'do nothing' regulatory scenario in which good firms do what they can to prevent or avoid causing environmental issues, but are undercut in the market by firms saving money at the expense of the environment and the affected communities. This can happen where there is no clear guidance from the regulator, lax observance of the law, no requirement or process to comply with any guidance and no communication of what compliance looks like, for example through a training programme.

Option	Success framework in place?	Enforcement used?
Hands off	✓	✗

In the absence of enforcement, but with a good guideline and industry training used in the context of a regulatory framework, I believe that, over time, we'd see an improved standard of environmental performance on worksites – but it might take a long time, with valuable environmental resources being damaged along the way. It could also end up creating a tilted playing field for the industry, with responsible operators losing out on work if they include the cost of environmental protection measures in their pricing when less scrupulous operators do not. It might also leave the environmental regulator open to accusations from the community that it is not doing its duty to protect the environment.

Option	Success framework in place?	Enforcement used?
Fair play	✓	✓

Having gone through this exercise, we can see that a faster track to good environmental performance that is supported by a good regulatory framework (accompanied by a good guideline and industry training) and the appropriate use of enforcement to ensure that regulatory requirements are met together provide a faster track to environmental protection. Major organizations such as utilities and those with large procurement needs can also 'require compliance' from their suppliers. To quote Earl Shaver, we should 'educate before [we] regulate'.[29] If we take an industry partnership approach from the early scoping stages, once all elements of the programme are in place and industry training is well under way enforcement will support good practice among the responsible firms by penalizing those laggards who let the entire sector down.

✎ **ACTION PLANNER**

How do the ideas in Figure 3.4 and Table 3.4 stack up against your own instincts, experience and organizational approach? Use Action Sheet 3.10 in the free Action Planner that accompanies this book to list the pros and cons of regulation and enforcement as a means of incentivizing good workplace performance.

3.6 PROGRAMME MONITORING, EVALUATION AND REVIEW

You can't manage what you don't measure.

Various attributions[30]

If you consider outcomes, objectives, monitoring, evaluation and review *up-front* in your programme planning, then your programme is much more likely to be robust and effective.

It's disturbing how many environmental management programmes are set up in such a way that it's not possible to evaluate them. I've scrutinized several, and it's heart-breaking to tell dedicated and hard-working professionals that the effectiveness of their programme can't be fully evaluated because of fundamental flaws in the structure of their plans and a lack of the right kind of monitoring data.

We'll look at how to evaluate the effectiveness of your *training* in Chapter 6. Here, I focus on how to evaluate that of the wider *environmental management programme* of which your training is just one part.

Evidence-based policy is the zeitgeist. It's impossible to evaluate the effectiveness of any programme if you don't build in your objectives and indicators – qualitative and quantitative – at the very start. Once you've done this, then managing your training within your programme will be much easier.

Nearly 30 years ago, I found a print copy of a journal article about a meta-analysis of the effectiveness of more than 600 environmental education programmes. The article itself is long since lost in the mists of time and I've been unable to find it online. But two things I remember with crystal clarity are that very few of the programmes were set up with evaluation in mind – and that those few that were so set up were later deemed to have been more successful.

I've been fascinated by this ever since – and I believe it's no coincidence: by definition, simply thinking ahead – right from the start – about how to assess the effectiveness of your programme sets you up for success. You don't have to be an evaluation expert to begin with – but if you begin by thinking about evaluation, you'll be an expert by the end.

How much will it cost?

How much time and money to invest in monitoring and evaluation varies between organizations and their function, but for the public relations (PR) sector, Katie Delahaye Paine suggests a figure of 5–10 per cent of project budgets.[31] This won't cover the cost of environmental sampling, but it will help you to budget for the outcome evaluation model set out here.

An evaluation for every purpose

Evaluating the effectiveness of environmental programmes is difficult because the results may take many years to become evident and it may be difficult to work out what's causing any changes you observe.

However, there is a whole population of professional evaluators out there and they really know their stuff. Rachael Trotman is a professional evaluator who specializes in sustainability. She describes the following aspects of review or evaluation, all of which may be performed with your partners in a collaborative way at different stages of a project:

- needs assessment;

- monitoring;

- formative evaluation;

- process evaluation;

- impact or outcome evaluation; and

- summative evaluation.[32]

For the purposes of programme evaluation, your needs assessment tells you about issues and desired outcomes, while your monitoring, or research, gathers information to inform the different forms of evaluation you can use.

Monitoring provides the information we need to: identify issues, baselines and benchmarks; report on progress towards desired outcomes; demonstrate accountability; and inform good decision-making. It also enables us to assess the effectiveness of our policies, regulations and other methods. It's an essential part of adaptive management and continual learning and improvement.

Formative evaluations answer questions about how to improve and refine a developing and ongoing programme. Usually undertaken during the establishment phase of a project, they can also assess the activities of an established programme. Formative evaluations may include process and impact studies.

Process evaluations help you to understand and document how you are implementing your programme. They answer questions about the types and quantities of services delivered, the beneficiaries of those services, the resources used to deliver the services, the practical problems encountered and the ways in which such problems were resolved. Process evaluations help us to understand how programme impacts and outcomes were achieved, so that we can replicate effective programmes. They are usually done where innovative service delivery models are involved or where the technology and the feasibility of implementation are not well known.

Impact, or outcome, evaluations assess the effectiveness of a programme in meeting its measurable objectives and targets. They focus on what happened to programme participants and how much difference the programme made to the issues of interest. They are undertaken when it is important to know how well the objectives for a programme were met or when a programme is an innovative model whose effectiveness has not yet been demonstrated.

Summative evaluations answer questions about programme quality and impact for the purposes of accountability and decision-making. They are conducted at the end of a project or programme, or after a specified period of time for an ongoing programme, and usually include a synthesis of process and impact or outcome evaluation components.

 ACTION PLANNER

Use Action Sheet 3.11 in the free Action Planner that accompanies this book to record your notes as you consider what forms of programme evaluation will work best for your environmental training programme and to take notes as you look for environmental or other programme evaluators to support you.

Logic models: measuring your programme outcomes and testing your theory of change

Start your monitoring and evaluation off on the right foot by looking at:

- logic models, which allow you to map your programme, engage with stakeholders and test your assumptions – all on one sheet of paper;

- Professor Stephen Olsen's 'orders of outcomes' framework, which will enable you to assess your programme's effectiveness in the short, medium and longer terms; and

- the SMARTER checklist for framing measurable objectives.

Watershed researcher Ian Brown exemplifies the need for strong and simple conceptual frameworks that can help us to work out what to monitor:

> So far most evaluations of natural resource management projects have been formative because of the difficulties associated with measuring what are often long-term outcomes. Summative evaluation and review in natural resource management have been widely neglected, with a substantial gap between theory and practice.[33]

That's why logic models are so valuable: they encourage us to ask a few basic and important questions that help us to identify a small number of indicators to give a robust evaluation of our programme's effectiveness.

Programme logic, encapsulated in a logic model, is a way of designing programmes that ensures from the outset that their implementation and outcomes can be evaluated. It's a fabulous tool for setting up any programme or project. Defining the issues and setting measurable objectives is a core part of programme logic – and these must be firmly grounded in a strong policy case, backed up by your programme needs analysis and training needs assessment. Your partners and other stakeholders play an essential role in all this.

For programme planning, logic models focus on the intended outcomes, where the guiding questions change from 'What are we doing?' to 'What do we need to do?'[34]

1. What is the current situation that we want to change?

2. What will it look like when we achieve the desired situation or outcome?

3. What practices need to change for us to achieve that outcome?

4. What knowledge or skills do people need to change their practice?

5. What activities do we need to perform to foster their knowledge and skills?

6. What resources will we require to achieve the desired outcome?

7. How will we measure our progress towards the desired outcome?

The focus on ultimate outcomes encourages programme planners to think backwards through the logic model to identify how best to achieve the desired results. It helps to 'plan with the end in mind', rather than to consider only inputs (for example budgets, employees) or the tasks that must be done.[35]

To make sure that the environmental outcomes of your training programme are measurable, repeatedly check the language used in, and the linkages between, all of the programme's elements.

- When doing your initial research and framing your environmental issue, choose your words with care: they will already be leading you towards your choice of indicators for your programme monitoring and evaluation.

- Frame your objectives in terms of outcomes (desired end state), not actions (what you will do to reach it), using the SMARTER checklist at which we'll look shortly. Make sure you use the same words as those that you used in framing the issue.

- Select the methods you will use to achieve your outcomes, making sure they are the right ones to address the issues and achieve the objectives.

- Select your indicators and cross-check them against all of the above to make sure that the ones you choose will most cost-effectively measure progress towards or achievement of the outcomes and address the issues.

In terms of indicators, both quantitative and qualitative indicators are useful, including assessments of the actual effectiveness 'on the ground' of environmental training by way of comparative audits of sites at which people have or have not attended your workshops, as well as anecdotal information.

This approach will give your programme good internal logic and consistency, clearly linking objectives to issues, methods (actions) to objectives and indicators to all the above – all, hopefully, firmly grounded in your legal and/or community mandate (that is, the policy and business case and the training needs assessment) for doing the work.

Done well, logic models make it much easier to evaluate the effectiveness of our plans or programmes. We can test how well we applied the cause-and-effect theory to a particular issue (or issues) by thoroughly examining all contributing factors, including the internal plan logic, methods for implementing the plan, the resulting outcomes and any external factors that may contribute to or confound an expected outcome.

Programme logic also reminds us to keep asking the questions we are trying to answer as we select indicators. There is a wonderful example of these on the University of Wisconsin Extension website showing a logic model for a phosphorus-focused water quality programme that includes a series of programme evaluation questions.[36] It's a great model to follow.

The 'orders of outcomes' framework

There is a vast literature on the difficulties of cost-effective monitoring and integrating the results from different monitoring programmes within and among environmental agencies. Stephen Olsen developed his 'orders of outcomes' monitoring framework specifically to address the difficulty of monitoring interventions in complex estuarine environments.[37] I've used it for many years for that purpose, while I and my colleague Will Allen (who discovered Stephen's work) have also used it for plant and animal pest control and other purposes.[38] The framework has been adopted by the United Nations Environment Programme (UNEP).[39]

Will and I have built Stephen's framework into our logic models. We and our colleagues have used it with different groups of stakeholders, including planners, engineers, environmental scientists and communities, and have seen how it helps to clarify everyone's thinking and makes programme planning much simpler and monitoring more robust.

The 'orders of outcomes' framework defines the sequence and scope of institutional and practice changes needed to achieve the more visionary and longer-term goal of sustainable resource management in a way that allows them all to be tracked. This is why it works so well as part of programme planning: instead of using the language of inputs, outputs and outcomes, the framework deems all of these to be outcomes of the programme team's work.

Logic modelling assumes a good understanding of issues and causes. We all have our theories – often unconscious assumptions – of cause and effect, on which we set out the various interventions that we expect will achieve a set of expected outcomes.[40] As Figure 3.5 illustrates, 'if we do A, our target audience will do B, and then C will happen'. The logic model process prompts us to realize that we may be making unfounded assumptions about cause and effect, to look for and test these chains of assumed causation, and to make sure that every method is directed at the core issues.

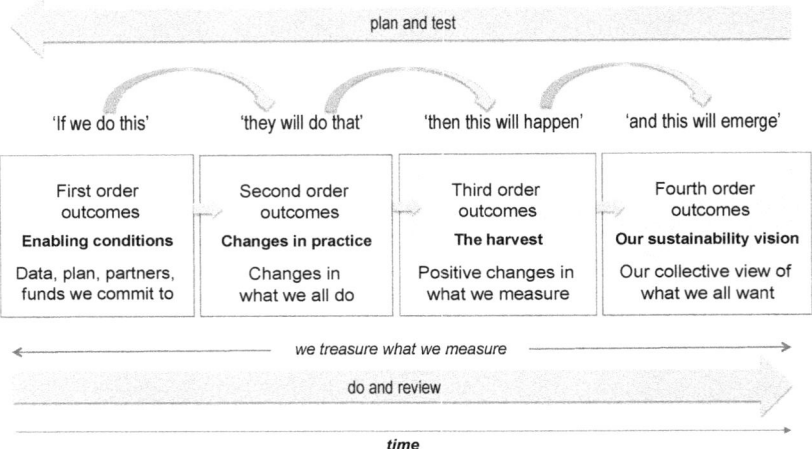

Figure 3.5 Clarifying and testing our theory of change with a logic model. *Source*: Adapted from United Nations Environment Programme/Global Programme of Action for the Protection of the Marine Environment from Land-based Activities (2006) *Ecosystem-based management: markers for assessing progress*. UNEP/GPA. Available at www.unenvironment.org/resources/report/ecosystems-based-management-markers-assessing-progress [accessed 5 June 2019].

In brief, the four orders of outcomes are:

- **first order outcomes** – the things that need to be in place at the outset, being enabling conditions such as funding, a robust plan, clear roles, goals and responsibilities, accountability, political commitment and stakeholder support;

- **second order outcomes** – evidence of changes in practice by ourselves and others through uptake of new practices and other forms of compliance with new regulatory and non-regulatory requirements;

- **third order outcomes** – changes in the indicators that measure environmental, social, indigenous and economic well-being; and

- **fourth order outcomes** – progress towards achieving the long-term vision of sustainable development, as expressed in an agreed vision.

One of the biggest difficulties for environmental programme evaluation is that many of the things we want to achieve – the third order outcomes – are long-term. But we can track progress by segmenting the time frames and documenting the first and second order outcomes intended to deliver them, as shown in Figure 3.6. These two – especially the observable changes in practice within the sector of interest – are very important. Making such estimates of likely time frames is part of the initial research for your own programme and the benefit of doing so is that it will help you to explain what's happening to key stakeholders, including the people who decide whether to fund your programme – and to sustain funding over successive electoral, directorial and managerial terms.

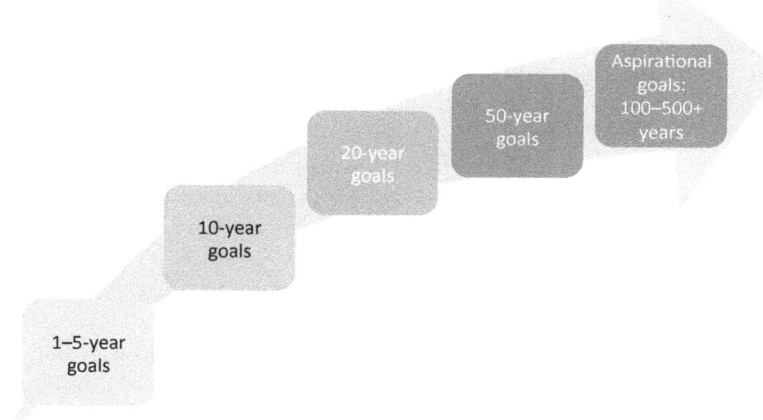

Figure 3.6 Using different time frames to maintain institutional momentum towards third order outcomes. *Source*: Adapted from Environment Waikato Lake Waikare and the Whangamarino Wetlands Catchment Management Plan Workshop of 27 March 2015.

It's this need for the long view that makes it so hard to gain sustained commitment to environmental outcomes that may take many years to emerge. Setting milestone targets that meet the goals of all parties will help you to demonstrate 'wins' along the way.

How long will it take you to achieve your outcomes? In Auckland, we saw second order outcomes, or changes in practice in erosion and sediment control, occurring right from the start of the programme: the training helped people to meet the new environmental standards in the new guideline and this change in practice was supported by feedback from council site inspections – and, after a grace period, enforcement for demonstrable laggards. There was a correspondingly prompt effect on third order outcomes for streams and coastal environments from the resulting avoidance of harm to water quality (here is where we see the value of baseline, or 'before', data).

However, Peter Mittermuller, a colleague involved with a construction and demolition waste reduction programme in Germany, told me it took 30 years for the new practices to be fully embedded on all building sites – that is, it was not until a whole new generation of workers arrived on site that environmental skills were found to have been included in their construction training.[41]

More cheeringly, for riparian restoration, some research suggests improvements in stream water quality and biodiversity can be seen in as little as two years, though results vary from stream to stream.[42] Similarly quick results were seen in the Whaingaroa Harbour – and, as well as the scenic and ecological benefits, a further benefit of ongoing riparian restoration has been the formation of some vibrant local businesses, showing once more how environmental initiatives create jobs.[43]

Over time, we become less confident about attributing outcomes directly to our plan, because external factors come into play; some may be synergizing (contributing to the outcomes we want) and others, confounding (negating or reducing the effectiveness of our actions). These factors are often beyond our control, but documenting them helps us to identify what influenced a given outcome. The 'orders of outcomes' framework thus provides a conceptual framework that helps us to tease out the many webs of causality at work in the complex environmental, social, indigenous and economic systems in which we work.

Using the SMARTER checklist to frame measurable objectives

The earliest use of the SMART acronym for objectives seems to have been by Peter Drucker in his 1954 book *Management by Objectives*.[44] I've adapted and added the final terms 'E' and 'R' from different expanded versions I've seen over the years:

- **S**pecific
- **M**easurable
- **A**ffordable
- **R**ealistic
- **T**ime-bound
- **E**ndorsed
- **R**elevant

Defining measurable objectives for environmental outcomes is 'challenging but achievable'.[45] The same could be said for setting up environmental monitoring programmes and integrating the monitoring of your environmental initiative into an existing monitoring programme.[46] There is a growing body of work on these topics, so ask for help from your environmental evaluators and monitoring experts.

As a quick tip, an objective or outcome with a verb in it is not generally a good one, especially if you say 'we will "start" doing X, Y or Z'. You are better off defining a desired end state. For example, the Waikato Regional Council's objective for maintaining the current water quality of Lake Taupo is stated as follows.[47]

The effects of nutrient discharges in the catchment are mitigated such that by 2080 the water quality of Lake Taupo is restored to its 2001 levels as indicated by:

Water quality characteristic	Mean	Standard deviation
Total Nitrogen (mg/m³)	70.3	19.1
Total Phosphorus (mg/m³)	5.57	1.4
Chlorophyll a (mg/m³)	1.18	0.6
Secchi depth (m)	14.6	2.7

 ACTION PLANNER

Use Action Sheet 3.12 in the free Action Planner that accompanies this book to help you to write SMARTER outcomes for your environmental training programme.

Programme review and evaluation: identifying the learnings

If you work your way through the order of outcomes, as illustrated in Figure 3.5, you'll find you are answering the questions we need to ask of any change programme.

- Where are we now? (Baseline)

- Where do we want to be? (Benchmark)

- How will we get there? (Methods/actions, including training)

- How will we know that we're getting there? (Indicators)

- How will we know when we've got there? (Objectives, milestones, targets)

We can track the progress of our programme by asking the following questions.[48]

- **What did our plan say we'd do?** A written plan is a first order outcome.

- **Did we do what we said we would? If not, why not?** We track two groups of actions: our own, in terms of how well we are implementing our plan; and those of others, in terms of whether the practice of our target audience is changing as we intended. Both of these are second order outcomes.

- **Did it make a difference?** Are the (third order) outcomes consistent with what we expected? Did the chosen interventions and how they were implemented influence these outcomes in the way expected? If so, this enables us to attribute at least a portion of those outcomes to interventions.

- **What else is going on that we need to know about?** What other trends are present or emerging? This refers to synergistic or confounding factors, which enhance or constrain the effectiveness of our interventions.

- **Does it make sense?** Given the monitoring data compared with our assumptions, how well do we understand the system? How accurate were our assumptions about cause and effect in selecting our methods in our plan? How well do we understand the environmental, social, indigenous and economic systems and processes that we are attempting to influence? This is part of reflection and review for adaptive management.

Monitoring these separate orders of outcomes yields the information we need to assess how well the plan is achieving or has achieved the desired third order outcomes in the short, medium and long terms. If the plan is not achieving the desired outcomes, the information helps managers and evaluators to differentiate between:

- **implementation failure** – that is, where expected outcomes are not achieved due to poor plan implementation, and

- **plan failure** – that is, where the plan's internal logic is flawed and the chosen methods are unable to achieve the expected outcomes, or the indicators selected are not the right ones to demonstrate the desired outcomes, or monitoring across all the outcomes is inadequate.[49]

In my experience, most plans fail because they don't invest enough time, money and effort in the first order enabling conditions and are often funded for periods of time that are too short to achieve third order outcomes. Most of the rest cannot defensibly be deemed successful because of lack of the right monitoring, typically because the programme managers focus on counting their own outputs (second order outcomes) and overlook the results the programme was set up to achieve (third order outcomes).

There is also what I call the 'out of left field' failure, where external confounding factors over which you have no control reduce the effectiveness of your interventions. For example, if you want people to reduce fuel use and therefore you run workshops on good driving and other green fleet practices, but the price of fuel suddenly drops (it's an unlikely scenario, I admit!), there is no longer any economic incentive for people to attend, despite the excellence of your stakeholder engagement, training and marketing.

Conversely, your training could be outrageously successful because a synergistic external factor crops up that has the effect of making everyone want to attend – skyrocketing fuel prices would mean our green fleet workshops were in high demand!

Figure 3.5 thus shows that our confidence in attributing outcomes directly to our interventions decreases across the orders of outcomes, as these external synergistic and confounding factors affect our progress towards our long-term outcomes.

Ecosystems – human, natural and financial – are complicated. Take care to consider the following factors.

- **Correlation vs causation**   Just because an indicator of interest has changed in step with another indicator doesn't necessarily mean that the latter caused the former: they may be separately responding to a third factor or totally unrelated. For environmental indicators, you need to carefully consider whether causation is present.

- **Attribution vs contribution**   If it seems that causation is present and that you can attribute a change in an environmental indicator to your actions and those of other parties, you also then need to work out how and how much those actions contributed to that change, given that many other factors may also be in play.

Now you can build a one-page logic model of your own programme.

 ACTION PLANNER

Use Action Sheet 3.13 in the free Action Planner that accompanies this book to build a logic model of your programme by revisiting your PEST and SWOT analyses to carry out a stakeholder analysis, by defining some fourth, third, second and first order outcomes, and by practising creating measurable (SMARTER) objectives for your first–third order outcomes.

Learning to love the learnings

Among the many other barriers to evaluating the effectiveness of our plans and reviewing them are lack of time for overworked staff to look back at their achievements and a widespread culture of fear in both public and private spheres. People are scared of 'getting it wrong'.

I suggested at one meeting that perhaps it could be a good idea to look at how well past plans had worked before hastening to prepare a bright and shiny new plan, only to be horrified at the response: 'Oh, we couldn't possibly do that – there'd be too much blood on the floor.'

This attitude is so sad. In both business and government, I see people who love their work. Giving them the time and tools to step back and reflect on their organization and its operations is part of the process of developing a form of reflexive organizational awareness and is a major step on the path to becoming a learning organization that 'gets' continuous improvement. This sets alight a love of learning about our work by way of monitoring, review and adaptive management – something environmental agencies talk about a great deal, but find harder to do in practice.

Monitoring and evaluation of environmental programmes demands a mix of holistic and strategic, yet targeted, approaches – especially if we are aiming to assess outcomes and effectiveness across the well-beings or capitals we are measuring (environmental, social, indigenous, natural, financial). Sometimes, minor tweaks to our methods will create benefits across several of these well-beings, and we need to track this thinking and doing and the outcomes that result.

At first, people worry that they can't think of anything to measure. Once they get into it, however, they find the reverse problem: they come up with more indicators than anyone would ever have the time and money to measure.

The solution: start small, focus on three or four leading indicators and progressively build your confidence over time. A small number of focused and measurable indicators is better than a large number of vague ones. Learn to see programme review as a tool of successful adaptive management. Here, I strongly recommend participatory or collaborative monitoring and evaluation.

Participatory or collaborative evaluation (depending on the degree of involvement of the parties) provides a powerful model for building partnership and knowledge. It is ideally suited to a capacity-building programme such as the environmental training based on partnership.

Some of the benefits of participatory or collaborative evaluation are that it's:

- more fun;

- likely to be more accurate and certain to be more holistic;

- more transparent, balanced and equitable;

- a way of encouraging people to reflect on the results of past actions and think more clearly about their future actions; and

- likely to promote better on-the-job support for the newly learned behaviours when the right parties are included.

By contrast, traditional forms of monitoring and evaluation can result in:

- measuring performance against pre-set indicators, often with the help of outside experts and not until the end of the project, so that mid-course corrections aren't possible and participants learn little about evaluation;

- too much concern with effort, effect and efficiency, and with the tangible and material performance of the project;

- a preoccupation with quantitative data over qualitative information;

- a bias towards favourable outcomes and a failure to capture unforeseen or unwanted consequences;

- a bias towards external conception and implementation, taking little note of the experiences of the local people or industry players with a major stake in training programmes;

- time-consuming major evaluation exercises that absorb people's energy;

- all stakeholders losing their focus on performance improvement, ongoing learning and adaptive management;

- an inability to learn from 'surprises'; and

- a fear of monitoring among project participants.[50]

It's never too late to start talking about participatory or collaborative evaluation – but it's even better to start the dialogue in the planning stages of your programme. Work with the results of the stakeholder analysis you did up-front in your logic model. Start with your internal partners and stakeholders, move on to your external partners and stakeholders (especially your industry focus group), and then ask for feedback from your trainees and their supervisors. It's also a good idea to ask your site inspectors or auditors to include questions about how useful the workshops have been in helping people to get things right on site and what other things could be included in the training to make it more effective. Questions about the availability of on-the-job support will also be helpful in identifying the factors contributing to good and poor practice alike.

This information will help you and your key partners to collaborate in evaluating – and improving – the effectiveness of the training. As Will Allen says, stakeholders don't all want the same level of participation.[51] There is no need to immediately involve reluctant stakeholders and stakeholders may change their level of involvement as the process continues, perhaps with initially reluctant parties coming on board or those whose concerns have been allayed departing. Your partnerships should thus be flexible and designed to grow.

Such participatory evaluation can make a major contribution to capacity-building and organizational – and indeed, I believe to industry-wide – learning. It recognizes the importance of all stakeholders evaluating the participatory process in which they are involved. For example, funders need evidence that their investments are paying off and that intermediate indicators of success (for example within the time frame of funding cycles) are being achieved; other stakeholders, who give their time to help (for example site project managers freeing up staff to attend training or hosting site visits) need evidence that their input is having an effect if they are to remain motivated to continue their involvement.

This requires clear objectives and indicators of success that promote accountability, and which can be monitored and evaluated by both participants and decision-makers.

The monitoring and evaluation component of environmental programmes needs to be as much about building capacity, diagnosing constraints and opportunities, and trying to make programmes grow and expand as it is about measuring and describing progress against targets.

Participatory processes encourage the use of evaluation as a learning tool and allow different team members to articulate their perspectives. They also provide information to feed into programme design, enabling the programme managers, in partnership with team members, to rethink goals and methods according to emerging issues.

Will Allen's paper concludes by noting that it's often useful to have a third party help with evaluation.[52] He quotes Carolyne Ashton, who points out that professional evaluators are not expected to have answers, but they are expected to raise important questions for participants to answer.[53] Ideally, they should specialize in helping the different parties to frame realistic goals, measuring progress towards operationalizing them, recognizing when a change of strategy may be required and extracting insights from their hard work.

Take the time to stop and think

Once you've done some monitoring, *take the time to stop and think*. We are all so busy that this is one of the things we don't often get round to, but, properly done, it is not only enjoyable, but also immensely valuable to take time to evaluate our environmental and training monitoring data and reflect on what it all means. A day retreat in a beautiful place can help.

Present conference papers. This book came out of two papers that I wasn't paid to write, but which gave me a priceless opportunity to collect information and capture my own thoughts and those of others, and to have to set them out clearly enough to be useful for other people. Jay Wilson said the same thing when he presented his 2011 paper on the City of Charlotte's erosion and sediment control programme. Going public in this way is also a big incentive for you to conduct an evaluation of the effectiveness of your training.

The process also made me aware how much organizational knowledge is stored in people's heads and how easy it is to lose it when they change roles or move on. Every programme in every organization needs its story regularly updated, so that we can see where we came from and how well we've done. What's more, this information can inspire future action: so often, we record our aspirations for a programme or we plan exciting new research or delivery, yet never do it. Revisiting our past can make sure these ideas are still ready and waiting for us when their time comes.

Sixty-two excuses for not evaluating the outcomes of your programme

You may have a good system for monitoring and evaluation – but do you actually do it? There are some obvious reasons why people don't do it, such as overlooking the need to build in time and money and, if necessary, the external skills to help you. But there are always plenty of other excuses. In 1994, I found a wonderful old UNESCO document, evidently collected by an evaluator who kept finding there was no data he could use.[54] Having had exactly this experience myself over many years of evaluating the effectiveness of watershed management plans, I can vouch for the fact that these excuses are still valid today.

The UNESCO team who publicized this list gave the following caution:

> As part of our untiring efforts to make life easier for overworked project directors, here are some of the more original responses to evaluation propositions, collected by a colleague in another organ of the United Nations System. When a single reason is not enough, one can always give two or three, for example, 'our project is different and the costs are too high'. However, one must be careful that the two reasons given do not contradict each other, for example, 'we are constantly evaluating' and 'our project is impossible to evaluate'.[55]

 ACTION PLANNER

Use Action Sheet 3.14 in the free Action Planner that accompanies this book to highlight your favourite excuses and add any more original ones of your own creation… then figure out what to do about them, so that you really do go ahead and measure your success and become a lifelong and lifewide learner!

3.7 PROGRAMME RESOURCING AND ENDORSEMENT

> Don't tell me what you value, show me your budget, and I'll tell you what you value.
>
> Former US Vice-President Joe Biden[56]

Developing and sustaining an environmental training programme is a big job and you will need help. Prepare your budget and action plan very carefully to make sure you can manage it well and maintain the support of your funders and other partners. No endorsement? No resources.

Resourcing

Your managers and elected representatives or board of directors will want to know what resources your training programme will need and how much it will all cost (see Chapter 4). Some costs are straightforward to assess; others are more difficult to estimate. The main categories of costs are transitional and operational.

Transitional costs are up-front costs that you won't incur again, such as:

- creating a guideline to establish the desired environmental performance standard;

- developing or adapting plans, policies and processes for regulation and/or compliance;

- developing training materials and training delivery mechanisms; and

- setting up administrative systems.

Operational costs comprise ongoing and intermittent costs, such as:

- administration, including managing registrations, venues and more (see Chapter 7);

- reporting, because ongoing communication and reporting within your organization or with your industry sector is vital to sustain your partnerships;

- marketing and delivering the training;

- carrying out site inspections to help to assess the effectiveness of your training;

- updating your guideline – a cost that can come up sooner than you predict, because both you and your partners and stakeholders will learn quickly and new technologies are constantly emerging;

- reviewing policies and procedures for cost-effectiveness;

- evaluating and reviewing the cost-effectiveness of the training and its wider programme;

- updating your training content and delivery; and

- making consequent changes to the programme.

Even if you run only half-a-dozen workshops a year, the organizational logistics can be daunting. Not all environmental managers have a personal assistant or access to an administrative support to help with this work. Moreover, especially in difficult times, government and other agencies may find it difficult to increase the complement of permanent staff, so you may need to consider other ways of getting help.

As an example, the up-front costs to develop or adapt a guideline include:

- research into available guidelines;

- engagement with your internal and external partners;

- staff time and/or consultancy fees;

- external peer review from technical experts;

- design and layout; and

- printing hard copies and/or developing web pages to promote the guideline and enable users to download it.

On the plus side of the ledger, consider the potential for:

- full or partial cost recovery of training, for example from registration fees or sponsorship (see Chapter 5);

- site inspections, if council staff or external auditors can charge their time to the project or site owner, for example; and

- financial inputs from internal and external stakeholders or in-kind contributions of time, venues, equipment and so on.

There is a growing move towards green financing, and regulatory bodies are increasingly using green bonds and other financial tools. Many philanthropic bodies also fund environmental initiatives, so it may pay for you to investigate local options. Remember too that you may be able to demonstrate the value of your training by monetizing the benefits – more on this in Chapter 6.

Few of us are lucky enough to have an open chequebook with which to achieve our big dreams. Many years ago, I worked with a cash-strapped organization and we came up with some principles to help it to work out what it could and should do first to deliver cost-effective environmental outcomes with limited resources. We analysed the existing situation and looked at the work that had already been done, then developed a vision for the future. We ran a series of workshops in which staff set priorities based on their principles and we came up with a five-year strategy, detailed action plans and a monitoring strategy that were all do-able with the limited time and money available.

It can sometimes be better not to start at all than to be unrealistically ambitious and start a programme that falls over, because this can engender disappointment and cynicism both within your organization and among your stakeholders. Carry on thinking and talking, and the right time may come along.

 ACTION PLANNER

Use Action Sheet 3.15 in the free Action Planner that accompanies this book to start listing the resources you will need for each of the elements of the Success Framework in your environmental training programme, considering your own operating constraints.

Endorsement

I have seen successful training programmes falter or fall over because they were taken for granted and had their resourcing cut or cancelled. So while your biggest and hardest sell may be when you're selling your business case to your managers, be aware that you are *always* selling the case for your training, especially when some of the outcomes won't fully emerge for some time. This can be as simple as:

- regularly and publicly thanking and giving credit to your external and internal partners, champions, volunteers and trainees;

- communicating success stories to them, the wider industry, the community and, of course, your funders – be they your senior management and board of directors, any external sources of funding or your elected representatives; and

- presenting regular reports to your funders on achievements against milestones and tracking resources, actions by you and others and outcomes (your first, second and third order outcomes), as set out in your logic model.

A colleague who, in a former life, had managed donations to a major public institution told me once that, when he stepped into the new role, he had to collate his own list of who the major donors were and discovered, to his horror, that these people – some of whom had donated millions of dollars – had never once been formally thanked! He was humbled to find that they were surprised and grateful to receive an official thank you, and was at a loss to understand why they had continued their donations without this simple recognition. To avoid this pitfall, you may want to prepare a communications plan to remind you who wants to know about your training programme, why you need to keep telling them about it, what they will want to know, how you will tell them, and how often you will tell them.

3.8   WHAT DOES SUCCESS LOOK LIKE?

> The most important single ingredient in the formula of success is knowing how to get along with people.
>
> Theodore Roosevelt[57]

Your business case will predict the successful outcomes you want your environmental training programme to deliver. But be on the lookout for those unexpected things that tell you your programme has been a success: they've taken me by surprise many times.

Here, I'll look at some of the indicators of the success of training programmes I've been involved with. Although I generally prefer to use the term 'effectiveness', 'success' is something in which we can more readily rejoice.

Emergence of a new profession

The most unexpected success of Auckland's erosion and sediment control programme was that, in bridging the gap between regulator and regulated, we ended up building a totally new profession: environmental and sustainability managers on large civil construction sites. Over time, these people have developed a level of expertise and collegiality that transcends the boundaries between regulator and regulated.

At first, the expertise was concentrated in the council, which had done so much work to develop the first erosion and sediment control guideline. Next, it spread to the consulting companies, which had to work out how to design the measures for their clients' construction sites. Then, it spread to the contracting companies as their supervisory and hands-on staff built their understanding and practical skills. Some years later, then programme manager Roger Bannister said he could be in a meeting with a developer, a consultant, a contractor and a council officer, and they would all know that they were part of the same industry – they were all working towards more sustainable development – and they would come together in conferences, workshops, field days, research and awards events.

This emergence resulted in a free movement of people and expertise between organizations, from regulatory bodies to consultancies to contractors and developers, and back again. The often-unanticipated benefits to all parties have been remarkable.

- First and foremost, the environment has benefited from ever-improving levels of protection.

- Regulatory processes have become more efficient as both sides have come to understand the other's point of view and have continued to share and develop their expertise, including by way of the movement of people between regulatory agencies and consulting and contracting firms.

- People moving between regulatory and commercial bodies bring with them knowledge and perspectives from their previous work. They so well understand the 'other side of the fence' that the fence almost disappears: at one recent conference, three papers were jointly presented by people working for a government regulator and a contractor – and, in every case, the pair concerned had previously worked for the 'other side'.

This movement of people among all the government, research, consulting and contracting agencies has been immensely beneficial. In the future, I'd like to see it include the tertiary learning sector and professional, educational, community and indigenous groups – for this would confer benefits even more widely still.

Better relationships

Another marker of the success of the training programme in the context of the Success Framework's partnership approach was the improved working relationships between the industry and the council, evidenced by more open communication and better mutual understanding. Remarkably, this was despite the council's sometimes vigorous use of enforcement mechanisms. The good performers were and remain keen to see industry laggards brought up to standard or put out of business if they consistently price jobs too low to include good erosion and sediment control. Similarly, they take it on the chin if their own performance has been less than excellent and promptly invest in their own in-house training to bring it up to scratch.

Uptake into tender specifications

After some years, we saw that the better environmental outcomes resulting from the training were taken seriously by agencies requesting tenders for public and private development and infrastructure projects, who started specifying that key people on each site must have attended erosion and sediment control training. Later on, they also started requesting that companies include in their non-price tender attributes their environmental track record and performance, as indicated by site scores, making it more difficult for poor performers to win work. As the sector has become more sophisticated, this has transformed into increasingly detailed environmental, social and indigenous performance requirements linked to carbon reporting and more.

Innovation

As the sector developed and extended its technical expertise, better ways of doing things emerged. Technical innovations emerged to deal with Auckland's difficult clay soils, including a low-tech chemical flocculation system. Other innovations followed, with the result that an assessment process was developed to manage these and a major review of the guideline was developed to accommodate the many improvements to existing controls.

Process innovations also emerged, such as the simple scoring system for compliance monitoring of erosion and sediment controls that formed the basis for the 'good environmental track record' so valued by the industry – and also for the council's initiation of enforcement action when needed. Other process innovations included the development of more flexible permitting provisions to reflect the fast-changing nature of development sites.

Improved productivity

Higher environmental compliance standards get a bad rap only from those businesses that don't know how they spur innovation and productivity – an effect known since the 1990s.[58] Consider the extreme example in Chapter 2, where a company subject to multiple prosecutions reluctantly rolled out environmental training, but did it so well that, within four years, its turnover tripled. The bewildered chief executive officer (CEO) couldn't understand how just one thing, the environmental training, could deliver that outcome. But the benefits of good environmental management flow into all aspects of businesses that embrace it as a source of competitive advantage. For some sectors, it also means getting more work from risk-averse government bodies, which may be the only source of work in hard times.

The business benefits of a good environmental track record are clear: a good record makes it more likely for tricky environmental permits to be granted to responsible operators, for example when applying for approvals to work in difficult seasons or sensitive locations. It's also widely recognized that environmentally responsible firms operate more efficiently on site and are less likely to run into enforcement issues, which cause expensive project delays and other contractual problems for the client.

The emergence of awards

As excellent practice emerged, the council decided to set up a system of environmental awards for the civil construction sector. Such awards are much more common now, but, in the early days of the erosion and sediment control training programme, they were virtually non-existent. The competition to win these awards and gain the subsequent reputational advantages meant another sign of success was seeing these awards being fully taken over by the industry.

A transferable model

As other parts of the council and other councils around the country saw the success of the erosion and sediment control training programme, they also adopted industry training to address other issues, some related to land development and others related to ecological enhancement, on-site effluent disposal, hazardous waste and more. There is enormous opportunity to use this model for other areas, such as industry-specific pollution prevention, rural land management issues, solid waste reduction and so much more.

A sustainable model

Impressively, the erosion and sediment control training programme survived several restructurings of the council, which, over time, have split up the previously integrated functions of research, policy, regulation, enforcement and education (including training), and have integrated eight separate councils in Auckland into one single one. After a lapse during and after the amalgamation, the training has made it back onto the new council's agenda and this in itself is another longer-term measure of success.

Better environmental outcomes

Additionally, of course, the original objective of the programme – to prevent and minimize sediment runoff from earthworks sites into infrastructure, fresh and coastal waters – was resoundingly achieved. Monitoring of site performance, water quality, aquatic biodiversity and community complaints have all demonstrated and continue to demonstrate the programme's success.

Supporting success: contributing factors

All of these indicators tell us how Auckland's erosion and sediment control training programme was successful – but what were the factors that led to that success? Here, I unpack some of the deeper aspects of elements in the Success Framework: expert trainers; technical excellence and commitment; pragmatic administration; and the focus on site inspections, with instant feedback.

As what 'real' trainers call 'subject matter experts' (SMEs), my co-trainer Brian Handyside and I had to get up to speed with best practice training to make sure we could get the council's core messages across to our trainees. This meant that our training was credible to the industry because of both the technical expertise of the two trainers and our commitment to becoming ever better trainers. We both attended 'train the trainer' training and I joined the New Zealand Association of Training and Development, of which I'm still a member nearly 25 years later. From the start, we adopted an applied 'learning by doing' training model, with small workshops (of 20–25 people), practical fieldwork and site visits, and a level and content pitched for the audience. As external contractors, we debriefed every workshop with our council client to see how we could do better, and we arranged regular meetings and reviews of new research and policies, and workshop content and delivery, to ensure our training remained fresh, up to date and relevant to both the council's and the industry's needs.

The programme maintained a focus on technical excellence, and its managers, staff and consultants were all water and soil experts. This was a major factor in the programme's success: their understanding of and commitment to erosion and sediment control gave them an understanding not only of the environmental controls, but also of the wider industry context and what was needed to support companies using them on site.

A pragmatic and responsive administration earned the respect, trust and cooperation of the industry. Construction projects are, by nature, dynamic: sites change rapidly as works progress and standard legal controls that work for, say, a manufacturing plant or a farm don't work for construction. Because the council managers understood this, they were able to set up a system of legal environmental controls that allowed quick turnaround of changes to erosion and sediment control plans. The council can approve a preliminary erosion and sediment control plan when granting a permit for the works and the conditions are written to enable a fast approval process for the plan updates that are constantly needed on dynamic earthworks sites.

Workplace reinforcement of training is vital to ensure that new learning is applied, rehearsed and retained. The erosion and sediment control programme included weekly on-site compliance monitoring by independent consultants, providing ongoing opportunities for dialogue, debate and mutual learning. Even better, the inspections were carried out on a cost-recovery basis, meaning that companies paid the full cost of their own compliance monitoring at no cost to the taxpayer. These inspections meant that the programme resources remained focused at the grass roots – on the sites – so that controls could be improved, repaired or maintained before they deteriorated or failed. This proactive and time-sensitive approach made a big difference to the level of environmental protection in Auckland and other places where it's been introduced.

The Chesapeake experience

I've been so fascinated by the success of the Chesapeake Bay Program that I used it as an interactive case study for 15 years, so I was thrilled to recently hear Rich Batiuk list his six lessons learned over three decades:

- sustain a management structure for consensus decisions;

- develop and defend a strong scientific foundation;

- develop numeric goals, report, reassess;

- define clean water clearly and simply;

- monitor, report to the public – and repeat; and

- set up a system for accountability.[59]

While the Chesapeake Bay Program was an environmental management, rather than an environmental training, programme, these six lessons are congruent with the seven elements of the Success Framework.

Summing up

The main message is to be on the lookout for those unexpected things that tell you your programme has been a success: they may be anecdotes from your partners and trainees, changes in practice that seem unrelated, such as tender specifications, media stories or social media posts, and many other sources. Capture this information and include it in your programme monitoring and evaluation. As you become more confident in using the Success Framework, you can pass this knowledge on to others.

 ACTION PLANNER

Use Action Sheet 3.16 in the free Action Planner that accompanies this book to list those things that indicate how successful your existing training programme has been so far – and get inspired by what your future success could look like!

NOTES

1 I have not been able to find a primary source of Golub's Four Laws of Computerdom, but a list may be found at http://ceir-miles.de/~eric/kabuff/texte/murphy2.html [accessed 15 August 2019].

2 This quote has been used by Cory Booker, Al Gore and Hillary Clinton, among others, and while some sources think it's probably not an African proverb, others think it may be, though the primary source is not known. See Joel Goldberg (2016) It takes a village to determine the origins of an African proverb. *Goats and Soda*, 30 July. Available at www.npr.org/sections/goatsandsoda/2016/07/30/4879257 96/it-takes-a-village-to-determine-the-origins-of-an-african-proverb [accessed 15 August 2019].

3 Beryl Oldham is a Fellow of the Human Resources Institute of New Zealand (HRINZ), has over 30 years' experience in organizational learning and development and HR, and a Master of Business Studies Degree majoring in Human Resource Management from Massey University. She is New Zealand's only Certified ROI Professional® (CRP). Find out more at www.completelearning.co.nz/about-complete-learning-solutions/ [accessed 29 April 2019].

4 Find out more about Grant Crossett's work at www.linkedin.com/in/grant-crossett-bappmgt-mzim-0b870940/ [accessed 29 April 2019].

5 R. Batiuk (2019) A keynote presentation and question-and-answer session at the 2019 Stormwater Conference and Expo of Water New Zealand, Mo Apopo – Stormwater – The Next Generation, 1–3 May 2019, at the Grand Millennium, Hotel, Auckland, New Zealand.

6 New Zealand Productivity Commission (2013) *Towards better local regulation.* Available at www.productivity.govt.nz/inquiries/towards-better-local-regulation/ [accessed 1 November 2019].

7 E.H. Shaver and F.M. Piorko (1991) *Education in Delaware's soil and water management programme.* A paper written while Earl and Frank were both with the Delaware Department of Natural Resources and Environmental Control. Presented to the USEPA Specialty Conference, District 5, Chicago, IL.

8 Find out more about Enette Pauzé's inspiring approach to leadership, partnership and stewardship at https://level8leadership.com/about/ [accessed 28 April 2019].

9 (1) Pam Warhurst, CBE, is the founder and chair of voluntary gardening initiative Incredible Edible in Todmorden, West Yorkshire: see www.incredible-edible-todmorden.co.uk/home. Her 2012 TED Talk at www.ted.com/talks/pam_warhurst_how_we_can_eat_our_landscapes?language=en has been viewed more than 1 million times [both accessed 28 April 2019]. (2) Radio New Zealand (2019) Growing silverbeet and self-esteem in Whanganui. A radio article from Nine to Noon broadcast on 7 May. Available at www.radionz.co.nz/national/programmes/ninetonoon/audio/2018693899/growing-silverbeet-and-self-esteem-in-whanganui [accessed 28 April 2019].

10 Find out more about Project Twin Streams at www.projecttwinstreams.com/ [accessed 5 June 2019].

11 Terry Pratchett (1987) *Equal Rites*. Corgi New Edition, p. 80.

12 C. Feeney, J. Crawford and P. Kouwenhoven (2009) *Good ICMPs [integrated catchment management plans]: telling the story*. A training resource for the Auckland Regional Council stormwater action programme; New Zealand Ministry for the Environment (n.d.) Quality planning website. Available at http://qualityplanning.org.nz/node/610 [accessed 28 April 2019].

13 Sourced from www.ziglar.com/articles/if-you-aim-at-nothing-2/ [accessed 15 August 2019].

14 See the UN Sustainable Development Goals at www.undp.org/content/undp/en/home/sustainable-development-goals.html; the indicators can be found at https://unstats.un.org/sdgs/indicators/indicators-list/ and can be downloaded in pdf and Excel format [both accessed 22 April 2019].

15 David Gebler, ethics engagement and integrated education at Lockheed Martin. Personal communication with the author, 19 August 2019.

16 Graeme Ridley, former Auckland Regional Council industry training manager and now director of RidleyDunphy Environmental. Personal communication with the author, 14 August 2019.

17 United States Environmental Protection Agency (2005) *National management measures to control nonpoint source pollution from urban areas*. Publication Number EPA 841-B-05-004, November. Originally accessed on 22 July 2012 at www.water.epa.gov/polwaste/nps/urban/index.cfm, but no longer available online; see the updated resources at www.epa.gov/nps/urban-runoff-additional-resources [accessed 5 June 2019].

18 Both of these works were used to inform a report to the Auckland Council in June 2012. I am grateful to the Council for permission to use and adapt material from that report, which may be referenced as: Ridley Dunphy Environmental Limited (2012) *Update of Auckland Region erosion and sediment control guidelines: TP90 gap analysis and literature review*. A report to the Auckland Council, June 2012.

19 Personal communication with the author, while preparing Roger Bannister and Clare Feeney (2009) Success by collaboration: the Auckland Regional Council's Erosion and sediment control training programme. A paper presented at the 33rd International Association of Hydraulic Engineering & Research (IAHR) Biennial Congress, 9–14 August, Vancouver, BC.

20 Lawrence P. Leach, (2005) *Lean project management: eight principles for success – combining critical chain project management (CCPM) and lean tools to accelerate project results*. Advanced Projects Inc.

21 Gordon Anderson (2019) Spitting sparks over H&S audits. *Contractor*, May, pp. 60–61.

22 Personal communication with the author, October 2017, while conducting research for a client.

23 Sheri Winter (2015) Compliance training: critically important, too often an afterthought. Available at www.caveolearning.com/blog/compliance-training [accessed 15 August 2019].

24 New Zealand Productivity Commission (2014) *Regulatory institutions and practices: final report*. Available at www.productivity.govt.nz/assets/Documents/d1d7d3ce31/Final-report-Regulatory-institutions-and-practices-v2.pdf [accessed 1 November 2019].

25 Professor Malcolm Sparrow (2000) *The regulatory craft: Controlling risks, solving problems and managing compliance*. Council for Excellence in Government. Brookings Institution Press.

26 Ibid., p. 3.

27 Ibid.

28 See e.g. www.ecan.govt.nz/your-region/your-environment/monitoring-and-compliance/investigations/ [accessed 15 May 2019].

29 E.H. Shaver and F.M. Piorko (1991) *Education in Delaware's soil and water management program*. A paper written while Earl and Frank were both with the Delaware Department of Natural Resources and Environmental Control. Presented to the USEPA Specialty Conference, District 5, Chicago, IL.

30 Variously attributed to business gurus W. Edwards Deming and Peter Drucker, but it's unlikely that either of them said this for the reasons set out in Don Peppers (2018) Why 'you can't manage what you can't measure' is bad advice. *LinkedIn*, 31 August. Available at www.linkedin.com/pulse/why-you-cant-manage-what-measure-bad-advice-don-peppers/ [accessed 16 August 2019].

31 Karen Scherer (1999) Market the event – then evaluate your success says industry guru. *New Zealand Herald*, 11 February. Find out more about Katie Delahaye Paine's excellent work at http://painepublishing.com/blog-2/ [accessed 25 October 2019]. Katie specializes in measuring the effectiveness of strategies used in the world of PR research and evaluation, and her thinking is very helpful and accessible.

32 Rachael Trotman (2008) *Promoting good[ness]: a guide to evaluating programmes and projects*. Prepared for the Auckland Regional Council. Available at www.tepou.co.nz/assets/images/content/training_funding/tools-for-learning/files/Promoting%20Goodness.pdf [accessed 1 November 2019].

33 I.C. Brown (2006) A review of the effectiveness of environment farm plans and integrated catchment management programmes. Unpublished report prepared for the New Zealand Ministry of Agriculture and Forestry, June.

34 P.F. McCawley (1995) The logic model for program planning and evaluation. Available at www.d.umn.edu/~kgilbert/educ5165-731/Readings/The%20Logic%20Model.pdf [accessed 5 May 2019].

35 See https://en.wikipedia.org/wiki/Logic_model [accessed 5 May 2019].

36 This excellent model is available at https://fyi.extension.wisc.edu/programdevelopment/files/2016/03/WaterQualityProgram.pdf and you can download the full report from https://fyi.extension.wisc.edu/programdevelopment/files/2016/03/lmcourseall.pdf [both accessed 5 June 2019].

37 S.B. Olsen (2003). Frameworks and indicators for assessing progress in integrated coastal management initiatives. *Ocean & Coastal Management*, 46, pp. 347–361.

38 Dr Will Allen is a social researcher who specializes in sustainability. His website is a wonderful resource for all sustainability practitioners: www.learningforsustainability.net/ [accessed 5 June 2019].

39 United Nations Environment Programme/Global Programme of Action for the Protection of the Marine Environment from Land-based Activities (2006) *Ecosystem-based management: markers for assessing progress*. UNEP/GPA. Available at www.unenvironment.org/resources/report/ecosystems-based-management-markers-assessing-progress [accessed 5 June 2019].

40 M. Day, G. Mason, J. Crawford and P. Kouwenhoven (2009) *Evaluating the effectiveness of district and regional plans prepared under the RMA*. Planning Practice Guide No. 4, International Global Change Institute, p. 8.

41 Personal communication dating back to the late 1990s–early 2000s when construction and demolition waste minimization initiatives were just getting under way in New Zealand.

42 S. Parkyn and R. Davies-Colley (2003) Riparian management: how well are we doing? *Water and Atmosphere*, 11(4), 15–17.

43 To find out more about the inspiring Whaingaroa Harbour restoration project, see www.harbourcare.co.nz/ [accessed 5 June 2019].

44 Peter Drucker (1954) *Management by objectives*. Harper Row.

45 N. Norton, T. Snelder and H. Rouse (2010) *Technical and scientific considerations when setting measurable objectives and limits for water management*. A report prepared for the Ministry for the Environment by the National Institute of Water and Atmospheric Research (NIWA). NIWA Project MFE10503.

46 See e.g. an older, but excellent, report: R. Beanland and B. Huser (1999) *Integrated monitoring: a manual for practitioners*. Environment Waikato.

47 Waikato Regional Council (2001) *Waikato Regional Plan (online version)*. You can see this objective in ch. 3.10 of the Plan, available at the Council's website at www. waikatoregion.govt.nz/Council/Policy-and-plans/Rules-and-regulation/Regional-Plan/Waikato-Regional-Plan/3-Water-Module/310-Lake-Taupo-Catchment/3102-Objectives/ [accessed 5 May 2019].

48 C. Feeney and P. Gustafson (2010) *Integrating catchment and coastal management: a survey of local and international best practice*. Prepared by Parsons Brinckerhoff and Environment & Business Group for Auckland Regional Council. Auckland Regional Council Technical Report 2009/092.

49 S. Wood, J. Crawford, M. Krausse and C. Feeney (2010) *Long Bay structure plan monitoring framework*. A report prepared for the North Shore City Council by Landcare Research Ltd.

50 W.J. Allen, M. Kilvington and C. Horn (2002) *Using participatory and learning-based approaches for environmental management to help achieve constructive behaviour change*. Landcare Research Contract Report LC0102/057. Prepared for the New Zealand Ministry for the Environment.

51 Ibid.

52 Ibid.

53 C. Ashton (1998) *Strategic considerations in facilitative evaluation approaches*. The Action Evaluation Research Institute online conference, September. Originally accessed in January 2002 at www.aepro.org/inprint/conference/ashton.html, but no longer available online.

54 UNESCO Internal Oversight Service (1991) 62 (good) reasons for avoiding evaluation in the United Nations system. Originally accessed on 8 March 2004 at www.unesco.org/ios/eng/evaluation/tools/outil_11e.htm, but no longer available online.

55 Ibid.

56 Joe Biden (2013) *Joe Biden: show me your budget and I'll tell you what you value.* Statement made in the opening seconds of the video available at www.youtube.com/watch?v=vuLwjFmESrg [accessed 5 June 2019].

57 Widely attributed to Theodore Roosevelt: see www.goodreads.com/author/quotes/44567.Theodore_Roosevelt?page=4 [accessed 15 August 2019].

58 Michael E. Porter and C. van der Linde (1995) Green and competitive: ending the stalemate. *Harvard Business Review*, Sept–Oct. Available at https://hbr.org/1995/09/green-and-competitive-ending-the-stalemate [accessed 29 April 2019].

59 Rich Batiuk (2019) A keynote presentation and question-and-answer session at the 2019 Stormwater Conference and Expo of Water New Zealand, Mo Apopo – Stormwater – The Next Generation, 1–3 May 2019, at the Grand Millennium, Hotel, Auckland, New Zealand.

4

The business case for environmental training

If you think education is expensive, try ignorance.

Comedy writer Robert Orben[1]

The value of a strong business case is that, in clarifying the issues and solutions, it points to the indicators you will use to measure the effectiveness of your training, including an assessment of the full financial return on investment.

4.1   WHEN TRAINING IS NOT THE SOLUTION TO THE PROBLEM

There is always a well-known solution to any problem – neat, plausible, and wrong.

US satirist H.L. Mencken[2]

Be prepared at an early stage to discover that training is not the solution to the problem – or, at least, not as a first step.

The following is a summary of some of the elements of the Success Framework that indicate when training cannot provide a solution to the problem – at least, not until other matters have been addressed:

- when research has not yet shown that training is a viable way of solving the specific sector problem you face, for example because similar local or international attempts have failed or a causal association has not been credibly demonstrated between the activities of the sector and the environmental problem;

- when there is no defined performance benchmark or guide to best practice, so there's nothing to base the training on, or when guidelines are available, but they are badly written or poorly thought-out and give no clear or consistent advice for the sector to follow;

- when there is no policy or regulation to make it possible to require the sector to follow a guideline, in which case persuasion will allow you to work with the willing and you will also need to be able to use some form of coercion to bring the unwilling up to scratch; and/or

- the right infrastructure to support good practice, which is essential for some sectors, is not in place – for example, training people to minimize construction and demolition waste won't be very effective until the physical and commercial infrastructure for collection, storage, pricing and sale of saved materials is in place.[3]

Don't worry if this means that training doesn't seem to be a viable solution to your problem. You may still need to prepare a business case to address these elements and, when some of the cornerstone elements are in place, you can progress to the training if it is still needed: training is always just part of a bigger picture.

 ACTION PLANNER

Do you think, at this stage, that training is the solution to your environmental problem? Use Action Sheet 4.1 in the free Action Planner that accompanies this book to help you to work through the elements you need to consider.

4.2 DECIDING TO DELIVER THE TRAINING

> Employers appear to be looking for training which will be directly applicable to their own business and provide a good return on investment.
>
> Teletrac Navman and CCNZ[4]

Your manager says, 'Okay, so the industry does need some training. But does it have to be us doing it?' What do you say in response?

Assuming that training is looking like a good solution so far, your business case needs to look at the most cost-effective ways of delivering it. Your managers are likely to ask if anyone else can or should do the training, so you need to think ahead about this.

A 2019 survey of New Zealand's civil construction sector showed that two-thirds of firms were looking to increase their capability in water, wastewater, stormwater, roading and residential infrastructure, all of which demand extremely high social and environmental outcomes.[5] It found that these firms delivered 'a whopping level of training' on the job, with two-thirds of respondents saying that more than 50 per cent of their staff training takes place on sites. As noted in the epigraph at the start of this section, the authors said that employers 'appear to be looking for training which will be directly applicable to their own business and provide a good return on investment'. Interestingly, they then said that 'the chance of this occurring is much higher if training is delivered in house', but added that 'sufficient training support from the industry's largest clients – local and central government – is vital'.[6]

It's more a 'both/and' than an 'either/or' situation, clearly pointing to a partnership approach to environmental training in which government training focuses on the desired performance standards and wider environmental outcomes, while businesses focus on their internal systems and procedures for meeting these obligations.

By now, you will have explored whether existing university or trades educators, professional and trades associations, or commercial and other training providers are already delivering the training you need. If you find it doesn't fully meet your needs, you may have to research how long it would take to develop or adapt the training for you, who would deliver it and perhaps how much it will cost. That information informs your business case.

If you work for a **government body** and find there is no suitable training out there, people will ask why you should run an environmental training programme, especially if the wider taxpaying community will be funding it. There are, however, good reasons why government bodies can justify running or supporting environmental training programmes. The most common is that the required training is not available elsewhere, because:

- the need for training often arises out of your initial research into environmental pressures and impacts, and such issue- or locality-specific needs are often not addressed in formal qualifications or able to be met by other training providers;

- when inspecting issues on manufacturing, farming, construction and other sites, compliance staff develop specialized and applied knowledge not usually taught in the relevant university degree courses – most of the environmental staff I know have learned on the job from others with hard-earned experience; and

- specialized environmental knowledge is not usually taught in technical or trades training, or at senior levels of secondary education, yet these may be the highest educational attainment of many staff working 'at the coalface' in sectors that create adverse environmental effects.

An emerging option is the development of licensing or franchise models of training that generate income for government bodies. I have seen examples in which a larger body develops a comprehensive package and shares it with smaller government bodies and businesses for one-off or annual fees that enable them to recoup the development cost over time, cover the ongoing operating costs and eventually return a profit. One such example in development would yield more income than local taxes to fund a significant environmental improvement programme. In another case, a major publicly owned utility has significantly raised the environmental performance it expects of its service providers across a wide range of infrastructure sustainability indicators, including carbon reduction and reporting, solid waste minimization, biodiversity and water quality, and has set up an environmental training centre to help them to comply. This could, in time, become a more widely available commercial service.

Chapter 1 outlined the difficulty firms face in recruiting environmental and sustainability professionals with the skills they need and sourcing the training needed to address skills gaps. This indicates a clear gap that government bodies can help to plug. Other positive reasons why environmental training by regulatory agencies goes down well are that:

- the simple act of outreach and partnership engenders tremendous goodwill in key sectors and the wider community; and

- government-sponsored training can help smaller businesses to access environmental training if they don't have the specialized staff to develop and deliver it in-house.

For these reasons, there is a global trend of environmental regulatory agencies delivering or endorsing training that meets specific environmental needs for which training is not readily available elsewhere. As well as the United Nations Institute for Training and Research (UNITAR), the US Environmental Protection Agency (EPA), various counties in the United States and several of the state environmental protection agencies in Australia deliver environmental training.

The training is often developed by staff from these agencies, along with experts from within the sector concerned, research institutes and tertiary providers. Some of the training is delivered by government staff and some, by the growing numbers of external specialist environmental training providers.

If you work in the **business or not-for-profit sector**, you may find government-run training programmes or environmental consultancies to help you to deliver the training. If not, here are some other options.

- If you are a larger organization and want specific training delivered in-house, then locate all the resources you can and engage a professional trainer to help you to develop, deliver and evaluate your training.

- If you are a smaller organization and can't do it all yourself or if you see an opportunity for the sector as a whole, consider presenting your business case for sector-wide environmental training to your professional, trades or sector association.

 ACTION PLANNER

Who will deliver the training? Is it you or someone else? Would in-house training or public training, or a mix of them, work best for you? How can a partnership model help? Use Action Sheet 4.2 in the free Action Planner that accompanies this book to list your actions.

4.3 BUILDING AND SELLING YOUR BUSINESS CASE

> The cost of training is paid once. Value is realized every time that
> knowledge or skill is applied.
>
> Job Training Systems, Inc.[7]

**'Show me the money.' What will you say when your manager asks you to do this? There are
two main reasons for preparing a business case: to get the budget you need and to identify
the indicators you will monitor to measure its return on investment.**

Before you invest a lot of time for no return, it pays to ask a few hard questions.
If you don't, you can be sure that your manager and accountant will do so, long
before you get to put the proposition to your elected representatives or board
members, who will probably ask the toughest questions of all.

Preparing a business case will help you to collate the benefits, disadvantages,
costs and risks of the current situation and compare them with those of your
preferred future situation, so that executive management can decide whether
the project should go ahead.[8] The information in the Success Framework will
have laid some of the foundation for your business case, especially around
partnership.

You can build a business case at any stage of your training programme: before
you start, while you're up and running, if you want to expand its reach or when
it's up for review. Other things to research when proposing environmental training include the following.

- Why now?

- What's been tried in the past and how well did it work?

- What people and existing resources are available to help us?[9]

We tend to focus on the costs of the training and the costs of the harm it's aiming
to prevent or minimize. The tools now available also allow us to quantify a wide
range of benefits. This is the realm of return on investment (ROI) – specifically,
the ROI of training. Demonstrating a worthwhile ROI will build the credibility
of your case with your decision-makers.

All organizations, whether public, private or not-for-profit, have a core mandate that sets out the things they must do. Many have a much wider range of
additional things they can also do at their discretion. When considering the
business case for your training, keep in mind all the needs and opportunities:
you may find that you can achieve a number of other outcomes, sometimes
called 'side benefits or associated benefits', for little or no additional effort, so
keep your options open as you clarify your ideas and progressively tighten up
your scope.

This chapter is not a 'how to' guide on building a business case. There is a vast
literature on making a business case for training of any kind and a growing body
for environmental initiatives. If your organization has its own in-house business

case template, supplement it with the references in the Further resources supplied at the end of this book.

Once you've got a grip on the business case methodology, you need to find the data to put in it. The aim is the same, but the data will differ for government and business bodies doing training.

Government bodies

For government environmental agencies, the sources of data and evidence for your initial research that can justify setting up a training programme and monitoring its ongoing effectiveness can include things such as:

- the state of the environment monitoring results indicating an emerging or intensifying issue;

- public complaints or concerns about the issue and/or its current management;

- monitoring of environmental permits indicating poor compliance with operating conditions;

- high use of enforcement compared with other methods of promoting good performance;

- information from reviews of plans, policies, regulations and enforcement that indicate the relative effectiveness of different methods used to achieve desired outcomes or the need for other methods;

- industry requests for assistance;

- the introduction of new legislation, policies, strategies or guidelines that change the performance benchmark; and

- evidence from other jurisdictions about the cost-effectiveness of training as a solution to the problem.

As shown in Chapter 3, the issue must fall broadly within the relevant legal mandate of your agency and there must be a demonstrable cause-and-effect link between the environmental issues observed, outcomes desired and observable performance issues in the relevant sector that training could address.

Case study: what the City of Charlotte looked at to see whether training was the solution to its sedimentation problem

The following are some of the things the City of Charlotte in North Carolina looked at when working out whether training would be a good solution to its sediment runoff issue.[10] They show how training is really part of the bigger picture shown in the Success Framework.

- **The intensity of construction activity** How much construction activity is going on? Over how big an area? Over how long a time frame per project? All told, is there enough activity to warrant setting up a training programme?
- **Jurisdictional-specific issues** What particular resources are at risk from the environmental effects of construction? How many water bodies are valued or impaired? What threatened species are present? What ability do the natural systems have to handle the site-specific and cumulative impacts of development? What measures should be used to protect them, and what guidance can be offered for building and maintaining those measures? What changes in local ordinances or procedures are needed? What integration with those of adjoining cities, counties or states might be needed?
- **Compliance issues** How many environmental notices are issued? How serious are the infringements or offences and their effects? What do they say about particular trends in non-compliance: are people installing measures wrongly, failing to inspect them or maintaining them poorly?
- **Knowledge level of the target community** Jay Wilson, the project manager, said that people 'didn't open up a big site with the intention to pollute – but they thought silt fences alone were the be-all and end-all of best practice'. It was clear that the industry's skills with estimating and pricing quantities and building roads did not extend to erosion and sediment controls and that a very basic level of training could make a big difference. Other training needs needed to be assessed, such as dealing with paint and concrete issues, as well as erosion and sediment control.
- **Community sensitivity and level of control** The wider community was very pro-development, but they came to see the benefits of protecting the environment while developing. The project permitting system needed to address community concerns, especially where local groups were very active. An advantage was that the comparatively high population made it easier to fund a training programme.

All these factors were weighed up to assess the needs for and benefits of a training programme. How do these compare with or inform your own considerations?

Businesses

For businesses, the sources of data and evidence for your initial research that can help you to justify an environmental training programme can include:

- the risks and opportunities posed by the transition to a low-carbon economy;
- new laws, regulations or guidelines that lift environmental performance benchmarks;
- pressure from customers, suppliers, lenders, insurers and shareholders;
- publicly available data on environmental issues and community concerns;

- increasing community concern or environmental activism;
- high staff turnover, with exit interviews revealing a desire for more explicit environmental commitments and actions from the firm and/or more training;
- your own records of environmental incidents and near misses;
- the need to make measurable progress towards giving effect to company objectives, targets, policies, systems and procedures;
- key performance indicators (KPIs) for the environment for relevant staff;
- advances in sector-wide understanding and performance improvement; and
- your own or the company's vision for a better future.

Organizations in high-compliance sectors may also consider things such as:

- the costs of project delays and contractual problems arising from environmental issues;
- a loss of confidence by investors, lenders and insurers;
- a loss of 'social licence to operate' from concerned consumers and communities;
- poor environmental compliance monitoring results; and
- the risk of being prosecuted or subject to lesser means of legal enforcement and the reputational and other harm that would ensue.

Sustainability is increasingly delivering innovation. I'm aware of two smallish firms (each employing around 50 staff) that have set up creative or innovation teams to deliver sustainability-based product and process innovations and, in line with business guru Michael Porter's seminal paper for the *Harvard Business Review*,[11] others see tougher compliance standards as an incentive to use sustainability to become more competitive.

'If training is expensive, what's the cost of ignorance?'

Gerald Richards suggests we consider the cost of *not* doing the training when preparing an ROI and business case.[12] He advises us to consider the cost of employee downtime, management intervention, staff turnover, the consequences of people misunderstanding an instruction and other time-wasters that results from inadequate training. This will, he says, help us to frame the questions we need to ask to start quantifying these costs.

The ROI calculation then adds up these costs and compares them with the cost of the training and the reduction of the costs of ignorance to give us a return and a payback period. What I liked about Gerald's analysis was his emphasis on the cost of *not* providing the training rather than the expense of providing it.

'Which has the greater impact when you present to your Board?' he asks: 'We have saved \$XXX by training 100 employees, or the cost of training 100 employees was \$X?'[13]

The scope of the business case and justification for training has shifted focus in line with employment trends.[14] People change jobs much more frequently, often to seek work with an employer whose values they endorse. Offering training and professional development thus becomes a source of competitive advantage in a market in which skilled staff are at a premium. Moreover, US research by the American professional body for trainers[15] reveals multiple benefits of good training:

- new staff become productive more quickly with good onboarding training;

- higher compliance and reduced risk;

- lower staff turnover and higher staff engagement and productivity;

- companies that offer comprehensive training programmes have 218 per cent higher income per employee than companies without formalized training; and

- these companies also enjoy a 24 per cent higher profit margin than those that spend less on training.

It may be time-consuming, but it's possible to make a strong business case for environmental training.

Setting priorities

Everyone needs to work out their priorities for environmental training. Criteria for companies include:

- **legal compliance** – the first step is to resolve any areas in which you are breaking the law;

- **reducing costs** – this is the easiest way of improving the balance sheet, so look at high-cost and/or high-volume inputs and wastes;

- **increasing revenue** – tendering, procurement and marketing based on environmental attributes to win the customers you want;

- **hazardous wastes** – using harmless or less harmful substitutes makes a big contribution to the health and safety of your staff, as well as to the environment, and often cuts costs as well;

- **'low-hanging fruit'** – getting some early scores on the board is good for morale and puts some cash in the coffers to fund the next initiative (use Bob Willard's spreadsheets to help you to find significant net positive returns[16]);

- **'bang for your buck'** – work out the next steps based on the likely ROI of the options;

- **risks and opportunities** – innovations and other options will emerge from trusting all your staff and value chain to engage in discussions around risk and opportunity; and

- **vision** – some people are more motivated by the rewards of striving towards a great vision for environmental and wider social benefits than by dollar benefits, so allow for these to emerge as you go.

For government agencies, the criteria also include risk.

Figure 4.1 is a simple risk matrix that regulators can populate with locally specific environmental sensitivities and compliance data. It is useful for companies to be aware of this approach. It allows them to self-classify and make their own decisions.

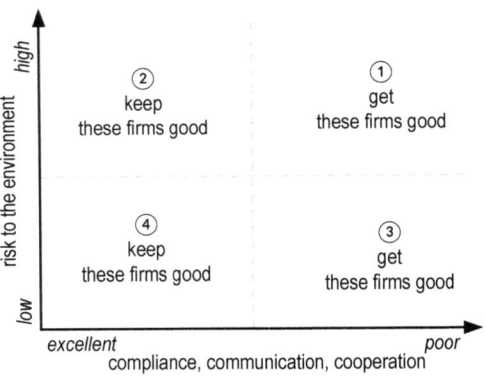

Figure 4.1 Simple risk matrix for prioritizing environmental training and compliance

 ACTION PLANNER

What information do you already have that could help you to build the business case for your environmental training? What further information will you need? How will you gather the information to build your business case that will be held in many heads and many documents throughout your organization? Who can help you? What other good resources can you locate? Use Action Sheet 4.3 in the free Action Planner that accompanies this book to list your actions.

Selling your business case

Getting your business case across the line needs a mix of reason and persuasion. Sustainability guru Bob Willard says that when executives push back on sustainability initiatives by asking overarching 'Why' questions, they are really seeking answers to what he calls 'the Big Three' questions of any value proposition – especially sustainability initiatives.

1. Is it the right thing to do?

2. Is it the financially rewarding thing to do?

3. Will bad things happen if we don't do it?'

You will need to answer these questions too. Bob says that if you answer 'Yes' to them, you may need to back up that answer with a detailed cost–benefit analysis and he lists what he calls the 'Ultimate 10 "Whys"' that you will need to investigate to do this. He summarizes these ten aspects of good business management in a simple, but powerful, diagram that is easy to follow.[17]

Facts are one thing; feelings are another. This is where persuasion comes in.

Changeology expert Les Robinson says that you need to wrap a 3-minute pitch around your vision for how to save the world and your extraordinarily cogent business case, so that if comes across as a 'safe, low-risk way to progress [your] managers' immediately salient goals. And [you] need a crisp, instantly comprehensible description that [your] manager can intuitively understand and sell up the chain'.[18] Les provides an eight-step process to guide you through the process and some invaluable tips. And it's not only brilliantly expert; it's very funny, too.

4.4 MEASURING THE BENEFITS ACROSS SEVEN FORMS OF CAPITAL

The global biodiversity crisis is so severe that brilliant scientists, political leaders, eco-warriors and religious gurus can no longer save us from ourselves. The military are powerless. But there may be one last hope for life on earth: accountants.

Jonathan Watts[19]

Monetizing natural, indigenous and social capital is a contested concept. But, in the indicator competition, the score still stands at 'GDP – 1/All Other Indicators – 0'. Let's get some scores on the board for a more balanced suite of indicators.

GDP – 1/All Other Indicators – 0

Gross domestic product (GDP) is the globally used indicator of economic growth. But it does not and cannot define goals that many people endorse, for example healthy, happy people and healthy built and natural environments. The perverse outcome is that environmental crises such as big spills make GDP look great, because it measures the money invested in emergency response and clean-up, while ignoring the costs of ecosystem harm and its local flow-on effects on fishing and amenity. This is because the economy has treated people and places as 'externalities' – that is, as being external to the business of commerce – with the result that neither governments nor businesses have been accounting for 'the damage their lawful activities inflict on nature and society'.[20]

This escalating damage has led Nobel prize-winning economists, accountants and scientists to conclude that 'we will continue to destroy the planet' 'unless we value the goods and services currently provided for free by the natural world and factor them into the global economic system'.[21]

Enter the six capitals: '[S]tocks of value that are increased, decreased or transformed through the activities and outputs of [an] organization. They are categorized ... as financial, manufactured, intellectual, human, social and relationship, and natural capital.'[22] The core concept is that the activities of governments and businesses should strive to add value to all the capitals and make full disclosure about trade-offs where a capital loses value by actions taken to enhance another.

Figure 4.2 shows the six capitals that smart companies measure. In this form, they model strong sustainability, where the economy is a subset of society and society exists within the natural environment. I have adapted this internationally recognized diagram by adding a seventh capital, indigenous capital, which – in many parts of the world – should also be addressed in order to reflect indigenous peoples' consistent contributions to sustainability.

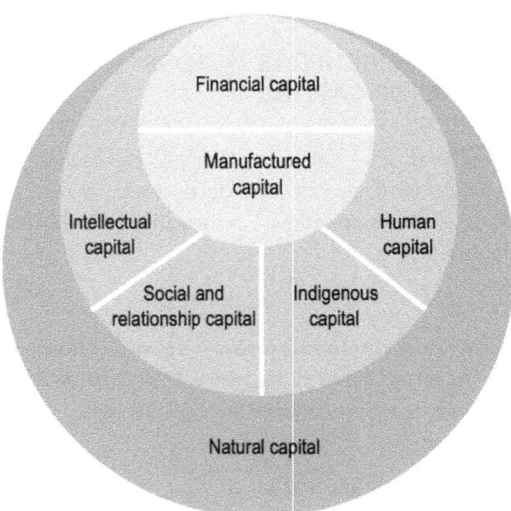

Figure 4.2 Strong sustainability at work: seven capitals for integrated reporting. Adapted (as suggested in §4.2, p.5 of the paper) from International Integrated Reporting Council (2013) Capitals background paper for <IR>. Available at https:// integratedreporting.org/wp-content/uploads/2013/03/IR-Background-Paper-Capitals. pdf [accessed 11 August 2019].

Every indigenous group will have their own definitions of the knowledge and values that inform their sustainability aspirations and practices and how they wish them to be applied. For the purposes of this book I define indigenous capital as a seventh capital, to and from which value can be added, in addition to the sometimes substantial stocks and flows of financial and other capitals that indigenous groups may have.

My perspective is that of a pakeha (person of Caucasian descent), informed by the Aotearoa New Zealand[23] experience, but the following terms and definitions, while specific to New Zealand, may serve as a starting point for discussions anywhere.

Indigenous capital may be partly described as Mātauranga Māori, a term that is often translated as 'Māori knowledge', or 'the body of knowledge originating from Māori ancestors, including the Māori world view and perspectives, Māori creativity and cultural practices.'[24] It can also be defined as 'the knowledge, comprehension, or understanding of everything visible and invisible existing in the universe'[25] and is often used synonymously with 'wisdom'. In the contemporary world, the definition is usually extended to include 'present–day, historic, local, and traditional knowledge; systems of knowledge transfer and storage; and the goals, aspirations and issues from an indigenous perspective' that help Māori populations develop frameworks for sustainably managing their own resources and also 'combining western and Māori perspectives to help solve' environmental and other problems.[26]

Like the other capitals, indigenous capital is not an add-on, by other parties, after the work has been done on the other six capitals, but must be assessed and developed by, or in partnership with, indigenous peoples themselves. It should also infuse and inform the approach to the other capitals.

In the more classically economic sense, much work is also being done around the world[27] on the financial capitals of indigenous peoples and First Nations, including their financial, manufactured and intellectual assets.

Businesses that report on the value they add across six or more capitals generally use the Global Reporting Initiative (GRI),[28] a detailed set of indicators that help them to understand, improve and communicate their impact on sustainability issues such as climate change, human rights, governance and social well-being. In their annual company reports, they also are encouraged to show how their operations are contributing to achievement of the United Nations Sustainable Development Goals (SDGs) (see Chapter 3). They generally do this by inserting short explanations and inserting the relevant SDG icon(s) in their reports where they can demonstrate that by adding value to each of the capitals, they are supporting progress towards the SDGs.

A growing number of stock exchanges around the world (getting on for 80, at my last count) have endorsed the six capitals. These stock exchanges now require their listed companies to provide integrated annual reports that address ecologically and socially, as well as fiscally, responsible management.

Governments can also report against the six capitals and align them with the SDGs, and can gather robust and accepted data on other well-beings, such as their national happiness and natural capital.[29] Such initiatives allow well-being indicators to 'compete with' GDP – or at least to complement it.

Environmental training adds significant value across all seven capitals that I espouse – adding indigenous capital to the usual six capitals[30] – thereby helping to make progress towards achieving the SDGs. Using such models, we can now design our environmental management and training programmes in a way that enables us to measure our local, regional and global progress across the full spectrum of internationally endorsed indicators of human and planetary well-being. I see huge potential for the capitals approach to attract the long-term funding we need for environmental training to really make a difference.

 ACTION PLANNER

Conceptual models of sustainability are important for helping us to understand complexity, organize our effort, and measure our outcomes to produce locally and globally consistent and comparable data. How could you use them to help you to set targets and make measurable progress towards global and local goals? Use Action Sheet 4.4 in the free Action Planner that accompanies this book to explore this.

The capital gains from restoring our waters

In the previous chapter, we looked at how to demonstrate alignment with espoused goals. It makes for a strong policy case – and a strong business case, too. What follows is a helicopter view of how a typical environmental restoration project might increase the value of the seven capitals. The definitions of the captitals come from the International Integrated Reporting Council (IIRC) and accountant extraordinaire Jane Gleeson-White.[31] My interest here is in training, but you can find out more about other activities in the sources from which these examples are taken.[32]

- **Natural capital** comprises the stocks and flows across nature's provisioning, regulating, habitat and other services. In July 2016, the Natural Capital Coalition launched the Natural Capital Protocol to provide a standardized framework that helps businesses to measure and value natural capital, to help them to benefit from understanding their relationships with nature.[33] Depending on the project's motivating issue, we would expect to demonstrate: higher capital values resulting from reduced frequency and severity of excessive flooding or pollution; improved biodiversity of endemic flora and fauna; and higher recreational and commercial harvests. People can rapidly become skilled in a range of restoration skills, with more than 70 people in one river basin receiving training in a range of skilled environmental tasks, sometimes gaining qualifications they may never otherwise have earned.

- **Intellectual capital** refers to education, training, skills, knowledge, ideas and intellectual property. Companies and communities taking part in environmental initiatives grow and exchange knowledge and skills and co-develop local solutions, gaining intellectual capital that will add value to their future activities.

- **Social, indigenous and relationship capitals** are collective capitals that lie in the connections and shared understandings among people, including (in the case studies cited) young urban Māori, helping them to work together for common purposes. One report on a stream restoration report estimated it needed training input, but delivered at least 21,549 hours of community education, and 84,278 hours of volunteer planting and weeding, with a lasting benefit to community cohesion shown by reduced littering and vandalism.

- **Human capital** is an individual capital that accrues to project participants. Seen as vital for sustainable development, it includes leadership, joy, passion, empathy and spirituality. In longer-term projects, local volunteers have the opportunity to become paid coordinators of community restoration efforts and, from there, they may move into other paid work as a result of learning a wide range of administrative and communication skills, often through vocational training and qualifications.

- **Manufactured capital**, sometimes called produced capital, comprises the built physical objects used to produce goods or provide services, including buildings, equipment and infrastructure such as roads, ports, bridges, and stormwater, water supply and wastewater systems. Many environmental restoration projects develop new manufactured capital in the form of walkways, cycle paths, bridges, bird-watching hides, interpretive signage, plant nurseries, fencing, park benches, art works, play spaces and more, often by giving on-the-job training to local people to do this.

- Local plant nurseries set up to support environmental restoration projects are a common example of **financial capital** gain. These often go on to be successful businesses that create new jobs by employing local people to help to meet the planting needs of farmers, councils, park managers, home owners and more. Some projects also generate increased income from increased passive recreation and tourism – all of which offer training and upskilling opportunities to local people.

In sum, environment and sustainability initiatives deliver significant gains across all capitals – if we take the trouble to measure them.

But what if your boss says that only your core environmental indicators count – and possibly some economic indicators of direct environmental harm – and

Table 4.1 Tracking outcomes for others to count: other benefits of stream enhancement measures. *Source:* C. Feeney, P. Kouwenhoven and J. Crawford (2008) ICMP *[integrated catchment management plan] review.* A report prepared for the Papakura District Council, December.

Item ID	Strategic outcome area / Description of measure	Growth	Infrastructure	Community involvement	Maori outcomes	Amenity values	Flooding	Other natural hazards	Water quality	Ecology
	Stream management: ecological measures									
STM 1.01	Riparian margins along all Category 1 streams shall be 10m from the edges of the stream bed on either side of the bank. Streambed width is taken as the mean annual flood extent.			✓	✓	✓			✓	✓
STM 1.02	Remove weeds and plant native plants at riparian margins. Undertake regular plant and animal pest control to maintain plantings.				✓	✓			✓	✓
STM 1.03	Install bio-engineering measures e.g. green gabions or bio-logs to protect the toes of streambank at risks of erosion.							✓	✓	✓
STM 1.04	Add meandering to engineered low-flow channels to mimic sinuous natural stream channels.					✓	✓		✓	✓
STM 1.05	Add logs, tree stumps and artificial eel holes and similar enhancements to increase habitat varieties at engineered channels.				✓					✓

that tracking social and indigenous indicators is someone else's responsibility? Yes, all outcomes should be counted, but perhaps others can count some of them for you. Table 4.1 is an example of a simple matrix developed to capture the many benefits of riparian planting in an urban area.[34] The waterways team weren't set up to measure all of them, but they were able to show the matrix to other teams in the council who were very happy to count these outcomes from their budget and report progress towards achieving their own targets resulting from the waterways team's work.

Why are you doing all this? Because it strengthens your business case. High-level goals, such as the SDGs and your national and state or regional policies, act as a compass to help you to find the direction your programme should be moving in. Remember that not all outcomes will be captured by hard metrics: the main idea is to deepen your understanding of the natural system and the people in it, and to work together to improve how we work.[35] High-level goals, together with your local plans and strategies, will provide you with a holistic set of relevant indicators to measure the outcomes your programme will deliver.

Now you can start to build your environmental training programme.

NOTES

1 Sometimes attributed to others, but most likely phrased as quoted by Robert Orben: see https://quoteinvestigator.com/2016/05/03/expense/ [accessed 15 August 2019].

2 H.L. (Henry Louis) Mencken (1920) *Prejudices: Second Series by H.L. Mencken.* Alfred A. Knopf, p. 155. See https://quoteinvestigator.com/2016/07/17/solution/ [accessed 15 August 2019].

3 Clare Feeney (2018) *ROI on environmental training: how to measure the full financial return on investment of your environment and sustainability training.* Workbook for a one-day workshop for environmental professionals.

4 Teletrac Navman (2019) *Construction industry report 2019.* Commissioned by Teletrac Navman and Civil Contractors New Zealand (CCNZ), p. 11. Available at www.teletracnavman.co.nz/lp/gc/confirmation?survey_results [accessed 9 August 2019].

5 Ibid.

6 Ibid.

7 Originally accessed in 2012 at www.jobtraining.com/trainingtips/twelve_thoughts. htm, but no longer available online. You can find out more about Job Training Systems Inc. at https://jobtraining.com.

8 M. Webster (n.d.) How to write a business case: 4 steps to a perfect business case template. Available at www.workfront.com/blog/how-to-write-a-business-case-4-steps-to-a-perfect-business-case-template [accessed 20 April 2019].

9 Michael Allen, with Richard Sites (2012) *Leaving ADDIE for SAM: an agile model for developing the best learning experiences.* American Society for Training and Development.

10 J. Wilson (2011) *Developing a local training programme.* A paper presented at EC-11, the 2011 conference of the International Erosion Control Association (IECA) in Orlando, FL, in February. Jay is a Certified Professional in Erosion and Sediment Control (CPESC) and works for the City of Charlotte in North Carolina. Find out more about the City of Charlotte's environmental activities at www.charlottenc. gov/StormWater/Pages/default.aspx [accessed 5 June 2019].

11 Michael E. Porter and C. van der Linde (1995) Green and competitive: ending the stalemate. *Harvard Business Review*, Sept–Oct 1995. Available at https://hbr. org/1995/09/green-and-competitive-ending-the-stalemate [accessed 29 April 2019].

12 Gerald Richards (2012) If training is expensive, what's the cost of ignorance? *Training and Development*, April.

13 Ibid.

14 V. Tran (2019) The business case for workplace training. *Training and Development*, Mar–Apr, pp. 47–49. References cited in that article include Third Sector (2017) *Workforce Trends 2017*. Available at www.thirdsector.com.au/107966-2; B. Phillips (2016) Is the 'job for life' mentality gone for good? Available at www.adzuna.com. au/blog/2016/02/04/is-the-job-for-life-mentality-gone-for-good/; Business Training Experts (2000) Profiting from learning: do firms' investments in training pay off? Available at https://businesstrainingexperts.com/knowledge-center/ training-roi/profiting-from-learning/ [all accessed 28 April 2019].

15 Dr Laurie Bassi (2000) *Profiting from learning: do firms' investments in education and training pay off? Investing in training improves financial success.* An executive summary of research conducted in 2000 for the American Society for Training and Development (ASTD), now the Association for Talent Development (ATD). Available with permission from ASTD at https://businesstrainingexperts.com/ knowledge-center/training-roi/profiting-from-learning/ [accessed 23 June 2019].

16 Bob Willard (2019) *Sustainability ROI workbook: building compelling cases for sustainability initiatives* Available at https://sustainabilityadvantage.com/books- dvds/roi-workbook/; see also his other books at https://sustainabilityadvantage. com/books-dvds/ [both accessed 20 April 2019].

17 Bob Willard (2016) *Three WHYs for any value proposition.* Available at https:// sustainabilityadvantage.com/2016/09/02/ultbook-answering-the-big- 3-whys/ [accessed 16 May 2019]. See also his other books at https:// sustainabilityadvantage.com/books-dvds/ [accessed 20 April 2019].

18 Les Robinson (2017) *How to pitch an innovation to a risk-averse manager.* Available at https://changeologyblog.wordpress.com/2017/02/14/grumpy-manager- innovation-pitch-updated/ [accessed 30 May 2019].

19 Jonathan Watts (2010) Are accountants the last hope for the world's ecosystems? *The Guardian*, 29 October. Available at www.theguardian.com/environment/2010/ oct/28/accountants-hope-ecosystems [accessed 29 April 2019].

20 Jane Gleeson-White (2015) *Six capitals – or can accountants save the planet? Rethinking capitalism for the twenty-first century.* W.W. Norton, p. xv. Find out more at www.janegleesonwhite.com/six [accessed 25 October 2019].

21 Pavan Sukdev (2010) *Global biodiversity outlook 3*. United Nations report. Available at www.cbd.int/doc/publications/gbo/gbo3-final-en.pdf [accessed 29 April 2019].

22 International Integrated Reporting Council (2013) *International integrated reporting framework*, p. 4. Available at http://integratedreporting.org/resource/international-ir-framework/ [accessed 7 April 2019].

23 Aotearoa means 'land of the long white cloud' – find out more at www.maori.com/aotearoa [accessed 20 October 2019].

24 Sourced from https://maoridictionary.co.nz/ (search for 'matauranga Maori') [accessed 20 October 2019].

25 Sourced from www.landcareresearch.co.nz/about/sustainability/voices/matauranga-maori/what-is-matauranga-maori [accessed 20 October 2019].

26 An interview with physicist Dr Ocean Mercier, winner of the Royal Society of New Zealand's 2019 Callaghan Award, broadcast on Radio New Zealand on 17 October 2019. Available at www.rnz.co.nz/national/programmes/ourchangingworld/audio/2018717818/a-bridge-between-science-and-matauranga-maori [accessed 18 October 2019].

27 See, for example, (1) BDO New Zealand Ltd 2019) The BDO Māori business survey report 2019. Available at www.bdo.nz/en-nz/industries/maori-business/maori-business-survey; (2) Chapman Tripp (2018) Te Ao Māori Trends And Insights March 2018. Available at www.chapmantripp.com/Publication%20PDFs/2018%20CT%20Te%20Ao%20Maori%20-%20English.pdf; and (3) KPMG (2018) Māui Rau: Aspiration to execution. Available at https://home.kpmg/nz/en/home/insights/2018/10/maui-rau-aspiration-to-execution.html [all accessed 20 October 2019]. Other work is being done by these and other firms, especially accounting and consulting firms, in Australia and Canada. I am most grateful to Troy Brockbank for his guidance on this topic. Any errors of interpretation are mine. Troy is Kaitohutohu Matua Taiao/Senior Environmental Consultant, WSP, Auckland, New Zealand, Beca Young Water Professional of the Year 2018 and a co-opted member of the Board of Water New Zealand.

28 The Global Reporting Initiative (GRI) helps businesses and governments worldwide to understand and communicate their impact on critical sustainability issues such as climate change, human rights, governance and social well-being. Find out more about the GRI Standard and download the monitoring indicators for free from www.globalreporting.org/standards [accessed 22 April 2019].

29 (1) J. Helliwell, R. Layard and J. Sachs (2019) *World happiness report 2019*. Sustainable Development Solutions Network. Available at www.worldhappiness.report/ed/2019/ [accessed 29 April 2019]. (2) The Natural Capital Protocol, a decision-making framework that enables organizations to identify, measure and value their direct and indirect impacts and dependencies on natural capital using an internationally standardized framework and tools. Available at https://naturalcapitalcoalition.org/natural-capital-protocol/ [accessed 29 April 2019].

30 International Integrated Reporting Council (2013) Capitals background paper for <IR>, p. 5 (the addition of other meaningful capitals helps to tell a 'unique value creation story'). Available at https://integratedreporting.org/wp-content/uploads/2013/03/IR-Background-Paper-Capitals.pdf [accessed 7 June 2019].

31 Jane Gleeson-White (2015) *Six capitals – or can accountants save the planet? Rethinking capitalism for the twenty-first century*. W.W. Norton. Find out more at www.janegleesonwhite.com/six [accessed 25 October 2019].

32 (1) See Whaingaroa Harbour Care at www.harbourcare.co.nz/ [accessed 7 June 2019]. (2) Morrison Low (2010) *Value case for Project Twin Streams*. A report prepared for Waitakere City Council in October 2010. Ref: 176103. Originally accessed in June 2012 at http://projecttwinstreams.com/, but no longer available online. (3) D. Buchan (2007) *Not just trees in the ground: the social and economic benefits of community-led conservation projects*. WWF–New Zealand, Wellington. Available at www.harbourcare.co.nz/wp-content/files/wwfnz_not_just_trees_in_ the_ground.pdf [accessed 29 April 2019].

33 Download the Natural Capital Protocol and helpful templates from https:// naturalcapitalcoalition.org/natural-capital-protocol/ [accessed 7 June 2019]. And if you think it's impossible or inappropriate to monetize natural capital, check it out: I'd expected to be daunted by the process, but was amazed at how lucid and systematic it is. Go on – try it!

34 C. Feeney, P. Kouwenhoven and J. Crawford (2008) *ICMP [Integrated catchment management plan] review*. A report prepared for the Papakura District Council, December.

35 Jed Emerson (2019) Exploring the purpose of you and your capital. Available at https://summit.boldentrepreneurship.com/talks/demystifying-the-purpose-of-capital-investment/ [accessed 12 September 2019]. See 25 mins in for the phrase 'the purpose of metrics is to inform practice'.

5

Building your environmental training programme

Environmental skills are increasingly among those in demand. This trend is likely to continue with 'green learning' consuming a larger proportion of corporate social responsibility budgets, and trainers who are knowledgeable about environmental matters and sustainability likely to be in greater demand.

Dr Brendan Moloney[1]

If we are to deliver excellent training, we need to understand how professional trainers go about their work. This global profession is as highly skilled as our own and the summary in this chapter is just a tiny glimpse into their knowledge. Join your local association of training and development to start your own wonderful training journey!

 TOOLBOX

Go to www.ESST.institute/Success/Toolbox to download some forms to help you through the ADDIE process, as well as many other useful templates for managing training workshops.

5.1 LEARNING HOW ADULTS LEARN

What we're seeing and needing is lifelong learning because of all the changes we see.

Josh Williams, Industry Training Federation chief executive[2]

Pedagogy is a deep skill, a high art and an increasingly robust science that environmental experts developing and delivering training need to grasp.

The term 'pedagogy' is that most widely used by professional trainers and instructional designers in Europe and the United States to describe the arts of teaching and training for children and adults. My favourite definition for adult learners is that of German scholar Jost Reischmann, who beautifully describes it as 'the science of the lifelong and lifewide education and learning of adults'.[3]

Pedagogy is about understanding how learning takes place and the principles and methods of enjoyable and effective instruction, as well as the orderly and logical arrangement of learning content into a series of topic steps.

Getting a professional trainer on board to build your pedagogical skills will be a great investment: good pedagogy is at the heart of good training. Shirley Caruso says:

> Adults typically want to choose what they want to learn, when they want to learn it, and how they want to learn; adults are able to contribute richness to class discussions and are considered valuable resources for learning from and with each other; whether or not an adult is ready to learn depends on what they need to know in order to deal with life situations; adults need to see the immediate application of learning; adults will seek learning opportunities due to some external motivators, but the more potent motivators are internal; and adults need to know why they need to learn something.[4]

What this means for trainers, says Robert Bogue, is that we 'have to realize that the way adults consume information and create knowledge is changing'.[5] This has big implications for how we set up our environmental training.

In the Further resources at the end of the book, you'll find a list of only some of the professional training associations around the world. It will be well worth joining one near you – or at least signing up to its mailing list so that you can find out about events of particular interest.

5.2 INTRODUCING ADDIE

> Whether you have 30 employees or 300, creating a culture of learning opportunity at your business will make a huge difference for your staff. Learning won't be restricted to set training periods, but will happen in all areas of your business, all day long.
>
> Sir Richard Branson[6]

When you know a lot about a topic, it can be difficult to distil out of an enormous volume of expert knowledge the small amount of key information that will help people to bridge a performance gap. The ADDIE model provides a set of basic steps to help trainers to do that.

Half the staff questioned in a 2012 training survey felt their training wasn't going to help them to do their job better, and only 28 per cent of them thought the work training they got was linked to their career development and goals.[7]

I'm sure you want your environmental training to be more motivating and usable than that.

This is why professional trainers place so much emphasis on how people learn and how best to design training that meets legal and organizational goals. The most widely used training development framework is the ADDIE model, developed in the 1970s and shown in Figure 5.1.[8] It has been revised and contested for many years, but remains a good overview for people new to training. Its stages are best envisaged as iterative processes within and among each other.[9]

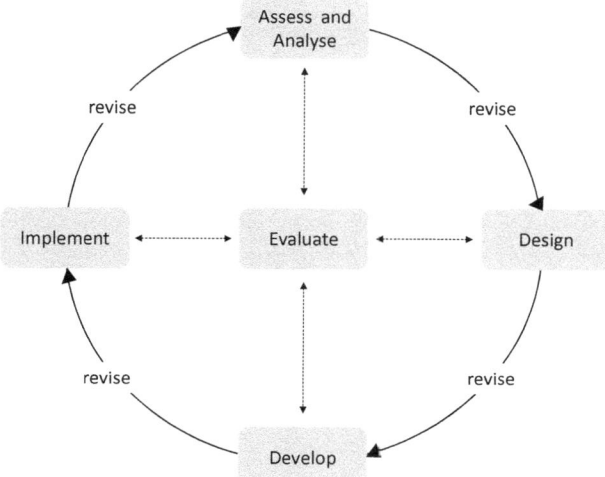

Figure 5.1 The ADDIE model of instructional design. *Source*: Eóghan Quigley (2018) ADDIE: 5 steps to build effective training programs. Available at www.learnupon.com/blog/addie-5-steps/ [accessed 16 May 2019].

The ADDIE stages are summarized in Table 5.1, based on online overviews supplemented by my many years of learning from outstanding professional trainers.[10]

In this chapter, we'll look at the first four elements of the ADDIE acronym. We'll focus on evaluating the effectiveness of your actual training in Chapter 6.

5.3 ADDIE: ASSESSMENT AND ANALYSIS – REVISITING TRAINING AS A SOLUTION

> The training effort in many organisations is often wasted as a result of poor (or non-existent) needs assessment.
>
> Change Factory[12]

Setting up an environmental training programme is a big, long-term undertaking. You will have answered some hard questions while preparing your business case; here, we'll drill down some more to check that training really is the best solution to your problems and, if so, precisely what the training needs to contain. This will also inform how you evaluate the effectiveness of your training, which we'll look at in Chapter 6.

Table 5.1 A summary of the ADDIE model of instructional design

ADDIE stage	Leading questions	Key outputs
Assess and Analyse		
A **needs assessment** identifies and prioritizes gaps in performance in terms of 'what *is*' versus 'what *should be*', while a **needs analysis** defines the causes of the performance gap and what needs to be addressed to bridge it.[1] At this stage, the needs analysis may identify that training is not the solution – or not the entire solution – to your environmental problem.	• What changes in practice are needed and who will make them? • Will training fully or partly achieve these changes? • Will these changes deliver the outcomes we need in practice?	A training plan that addresses who, what, when, where, why, how and how much A management plan to address issues that training alone can't solve
Design		
The **design** phase brings together the learning objectives, subject matter analysis, training content, structure and duration, lesson planning, practical and other interactive activities, delivery modes (for example online, classroom, on-site), who will be trained, who will deliver the training, when and where you will deliver the training, and what evaluation methods will be used to indicate whether the desired training outcomes are achieved. Seek input on this from your partners.	• What key aspects of adult learning apply to your audience? • How, when and where will they want to learn?	An overview of the pedagogical principles informing the training A clear design template or concept map of how it will be structured, delivered and evaluated, possibly including a blueprint, storyboard or prototype
Develop		
When you **develop** your detailed training content, you will do so within the design template or concept map, including workbooks and electronic components, guided by the core content and structure you've already defined. Make both your hard copy and electronic materials as succinct, attractive and easy to navigate as possible for your trainees. Check the basics: spelling, punctuation and grammar, and logical flow.	• Who should we ask to test our choice of structure, detailed content and training delivery modes to make sure the training suits the needs of our trainees and delivers our required outcomes?	Training materials in printed and/or electronic form A pilot training session that has generated helpful feedback from partners and stakeholders

Implement		
Once you've developed and piloted your training materials and processes, you can **implement** it – that is, actually deliver it to some trainees. If your organization has a learning management system (LMS), this will make it it much easier to manage registrations, book venues and trainers, send information out to trainees, and track attendance and completion of training and evaluation results. If it doesn't, set up a spreadsheet to track these things. Yes, it's a boring job – but, without it, you will never be able to report on your achievements and outcomes.	• How will you track your training? Using a spreadsheet? A customer relationship management (CRM) system? An LMS? • How will you review each workshop or electronic training module to see how effective your training has been and how you can review your design, development and delivery to keep doing better?	Feedback forms from the trainees who have attended your training indicating how effective it was – to help you to iterate back to review the assessment, analysis, design, development and implementation of your training

Evaluate		
In Chapter 3, we looked at the outcomes of the environmental management programme of which your training is a part and what indicators will help you to evaluate how effective it is. You will see that the ADDIE model builds in **formative evaluations** all the way through and, once you have repeated your training over the planned period of time, you will be able to do a **summative evaluation** before deciding what to do next. Asking open questions in your workshop evaluations will invite trainees to suggest things you might not have thought of.	• Did the training, as implemented, help us to address the performance gaps, objectives and outcomes we identified in the first two ADDIE stages and our business case? • If not, why not? • If yes, to what extent did we address them and how can we do even better? • Is any further training or other support needed?	A report to your managers, funders, partners and key stakeholders on your training evaluation results and what actions you will take in light of its findings

Sources: Adapted from (1) https://en.wikipedia.org/wiki/ADDIE_Model [accessed 16 May 2019]. (2) Eóghan Quigley (2018) ADDIE: 5 steps to build effective training programs. Available at www.learnupon.com/blog/addie-5-steps/ [accessed 16 May 2019]. (3) A series of presentations on ADDIE delivered in 2014 for the Auckland Branch of the New Zealand Association of Training and Development.

LEADING QUESTIONS

At the start of this stage, you are asking the following questions.

▶ What changes in practice are needed and who will make them?

▶ Will training fully or partly achieve these changes?

▶ Will these changes deliver the outcomes we need in practice?

Workplace performance and compliance issues generally fall into four categories of things that are missing or below par:

- knowledge – what people know;

- skills – what people do;

- motivation – their internal drivers; and

- workplace support – the built and cultural infrastructure to support new practices.[13]

Training alone can address only the first two of these, although environmental training can also improve people's motivation, either by inspiring them about how their everyday actions can help to heal the planet or by motivating them to avoid penalties. To build and sustain motivation, managers need support when stepping up to their new environmental coaching role: ongoing workplace support is critically important for the uptake of training.

Why don't people do what we want them to?

I really like Mager and Pipe's perspective on performance. They say that people don't perform as desired for many reasons – often because they:

- don't know what's expected;

- don't have the tools, space or authority;

- don't get feedback about the quality of their performance;

- don't know how to do it;

- are punished when they do it right;

- are rewarded when they do it wrong; and/or

- are ignored whether they do it right or wrong.[14]

The more we can talk in neutral terms, for example about a 'performance gap' rather than 'poor performance', the more open we can stay to the many factors contributing to that performance gap.

Assess the performance gap: knowledge and skills

Defining the performance gap is so important that there is a huge literature on it. Training can effectively address the knowledge or skills gap only if you can define the environmental performance gap by asking the following questions.

- What is the performance standard your trainees need to reach? This is the **benchmark** defined in your guideline or similar document.

- Who is it who needs to meet the standard? Exactly who is your target **audience**?

- What is their current **baseline** performance, as objectively observed and described? Exactly what is being done poorly? Who is performing well and how are they doing it?

- What is the **gap** between baseline and benchmark performance? What knowledge and skills will help trainees to bridge the gap? What specific and objectively observable actions do your trainees need to change or acquire? Is there a gap between knowing and doing, and if so, why?

Without this assessment, it will be very difficult to deliver targeted training as the solution to your environmental problem of concern. If your data is inadequate for this assessment, turn it into an opportunity to engage with the sector and other partners to start the whole process of applying the Success Framework.

You may find it useful to refer to Figure 5.2 below to help you to work through this process.

 ACTION PLANNER

Use Action Sheet 5.1 in the free Action Planner that accompanies this book to work out, as precisely as possible, who the target audience(s) are for your environmental training – the actual people whose routine work practices need to meet your performance benchmark.

Case study: great training for the wrong audience – not the best needs assessment

The following case study is the sort of embarrassing thing that can happen to trainers before they really appreciate how difficult it can be to identify the actual target audience for your training. One of New Zealand's major manufacturing companies, with many sites all over the country, wanted to update its environmental training so that its front-line site managers at the 'leading hand' level were kept up to date with higher expectations. The training objective was for the site managers to develop a personal action plan during the workshop and use it as the basis for implementing real environmental improvements at their respective sites.

I learned a hard lesson on this job about the vital importance of doing my own training needs assessment rather than taking the client's at face value. The first workshop went well. The trainees were most forgiving: they said it was a wonderful workshop, but that their managers should have been there, not them!

I'd begun to wonder during the day about how well the workshop was targeted: the trainees were clearly the most practical and capable people you could ever hope to have on your staff, but paperwork was not one of their strengths. Fortunately, my client was present and agreed that the 'personal action plan' was pitched at the wrong people. So the next phase of the training was amended to meet the actual on-site performance needs and immense practical skills of this target group – not those of their managers.

Not surprisingly, the subsequent nationwide series of workshops went much better. The new workshops focused directly on these site managers' legal environmental responsibilities and the standards, process and management behaviours expected of them. Part of this was about knowing the differences between 'good' and 'bad' practice: what did each look like? The company's history of good environmental management and regular audits meant that it had a wonderful file of 'good, bad and ugly' photos. We put the trainees in small groups and asked them to look at three or four of these photos. Each group had to report on what was 'good' and/or 'bad' about each photo, and why. This approach recognized and built on the trainees' skills and experience by encouraging them to work things out for themselves and to learn from each other. Lecture-style presentations don't work for any of us, let alone for competent, hands-on people!

The importance of the training was reinforced by showing a special video in which the chief executive officer (CEO) emphasized the value the company places on good environmental performance, and interviewed site managers from different sites about some of the company's particularly effective initiatives. This really captured people's attention.

Another attention-getter was presentations from invited local environmental regulators. It is so much more credible to have information about environmental issues and their significance and local enforcement policies and procedures delivered 'from the horse's mouth', rather than by the trainer.

Analyse the causes of the performance gap

The next step is to identify the *cause* of the performance gap. Is it something that training can solve? This raises more questions to ask, as Mager and Pipe suggest.[15] The performance gap may be a consequence of any of the following factors.

- **Knowledge and skills** If trainees' performance after training is still below par, does this indicate either that the training was inadequate or that trainees' performance is a people management, rather than a training content, issue?

- **Motivation** Can trainees' lack of internal motivation be addressed by ensuring that they fully understand why the training is needed? For example, a training outcome might be trainees saying, 'I didn't realize that stormwater pipes can carry pollutants into streams and the sea and harm life there. I like fishing – so I want to protect my enjoyment of food and recreation.'

- **Workplace support** Is there a lack of built infrastructure to support new practices and/or a lack of organizational motivation, in terms of workplace culture – in other words, unsupportive attitudes and actions with respect to the environment?

If workplace factors make it too difficult for our trainees to apply what they learn, then training will not solve our environmental problem. Here are some common examples of wider issues with the high-level support in a trainee's workplace that may render our training ineffective.

- Sometimes, systems or procedures don't allow new actions to occur. If, for example, local engineering codes and/or planning ordinances don't allow water-sensitive urban design, then training can't solve the problems associated with conventional development until they do.

- Infrastructure or equipment can be a barrier to new behaviours on a site because of the plant or site layout or built infrastructure, or because essential things aren't available on site or in the region, such as recycling bins, spill kits or other specialist environmental needs.

- There are instances in which the resources necessary to take action, such as staff time, responsibility or money, aren't available, for example:
  - there is no one with environmental results in their job description or key performance indicators (KPIs);
  - environment and sustainability managers don't have their own discretionary budgets (a barrier that's surprisingly common); or
  - tenderings to provide goods or services are too competitive and/or environmental outcomes are not valued in the process.

- Organizational culture can stifle the practice of new behaviours, for example if:
  - a culture of fear or criticism doesn't allow failure and analysis as part of learning;
  - there is a disconnect between espoused environmental policies and what happens on the ground;
  - set attitudes – a poor company culture – block good practice, sending signals such as 'That's not how we do things around here', which are often unspoken and rapidly infect new people or practices; or
  - supervisors are not trained, supported or available for on-job follow-up.

- Key people within and/or beyond the organization can even block progress, for a range of reasons.

Often, these things reflect wider issues within individual organizations, such as complex and disjointed policy, management, regulation and infrastructure. Identifying the barriers may help you to identify what additional forms of support your training programme will need to be effective, and this will inform the actions you and your partners should take with respect to the Success Framework.

Quality managers are taught to ask 'Why?' at least five times to be sure they have drilled down to the underlying cause of a problem, so carry on your analysis until you and your partners have all reached an agreed understanding of the apparent and the underlying causes. Getting to that understanding will tell you two things: whether training is the solution to the problem; and, if so, what other supportive factors are needed to make the training more effective.

Fatberg alert: how good is our needs analysis?

Will our training treat the underlying causes of a problem or merely its symptoms? The importance of a good training needs assessment is highlighted in an example given by learning and evaluation specialist Helen McPhun, which I've adapted in the following case study and illustrated as Figure 5.2.[16]

Case study: good needs analysis avoids unnecessary training costs

Helen McPhun's scenario is that oil in drains is costing a company $200,000 per year in compliance fines imposed by the city council. That problem defines the issue that needs to be fixed; the next steps are to identify what solution the business needs and to set a goal for it. The firm decided to set zero compliance fines per year as the business objective. Having defined the issue and the goal, the following steps are to analyse the causes of the performance gap to see what training and/or other solutions are needed.

Here begins the detective work, just as in any other compliance, lean manufacturing or resource efficiency project. We might discover that a blocked trap, filter, valve or drain is causing the problem by allowing waste oil to get into the drain instead of directing it to a detention device for removal. If so, the solution is a straightforward repair job.

If we find that all the waste management systems are in good repair and working well, the focus moves on to what people are doing: what performance standard will achieve the business objective? As Helen said, 'the performance objectives would obviously be not to put oil down the drain. Seems pretty simple!' However, this already is the performance standard and staff have been trained to follow it. At this point, Helen points out, it would have been easy for the human resources (HR) team to simply repeat the training on compliance requirements – but if staff already know them, what is the *real* cause of the problem?

Observation reveals that, no matter how careful the staff are, the method used to apply the oil is impossible to control, with the result that pools of oil end up on the floor and flow into the drain (causing, I would think, quite a safety hazard, as well as an environmental compliance problem). In a brainstorming session, staff say that a hand-held trigger spray is a solution that would also eliminate the use of oil altogether from one step in the process. Their solution is adopted. Not only does it totally eliminate the compliance fines, but also it saves the company hundreds of thousands of dollars a year in reduced oil use.

This is a brilliant example in which analysis of the cause of the performance gap indicated that training was not the solution to the problem. Had the HR team simply re-run the training on the old performance standard, Helen says that the result would have been a loss on investment in training and no improvement in performance – and the compliance fines would have kept mounting up.

As a result of this deeper analysis, the company now has a new performance standard on which to base training for new staff.

Now imagine that our new method of oil application is in place, but waste oil starts causing compliance problems again. We do a new needs assessment to identify who among the staff are not meeting the performance objective and why. Here, Helen lists lack of knowledge, skill, motivation, confidence or support as possible causes. Again, not all of these can be solved by training. However, if it is a knowledge, skill or confidence issue, then training can indeed resolve the problem. Otherwise, the issue needs to be resolved not by training, but by improved staff management.

In sum, Helen says that a training need can be identified only when an assessment has gone deep enough to uncover the true issues and causes and what is needed to address them. Once training objectives, or learning outcomes, can be defined, training can be developed, delivered and evaluated.[17]

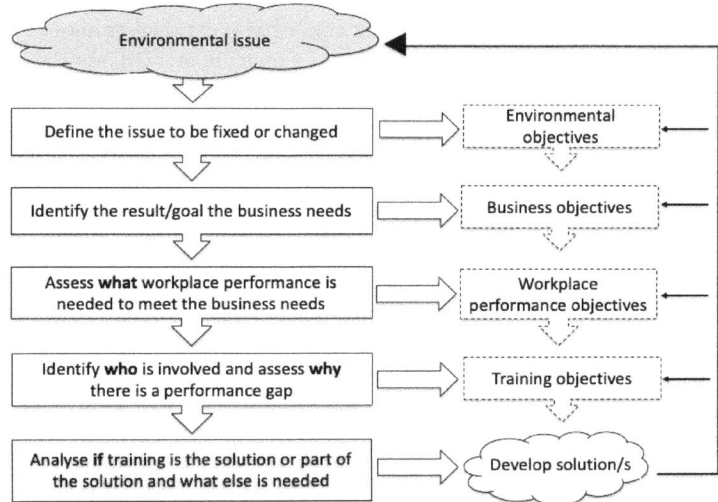

Figure 5.2 Needs assessment and analysis to align business, workplace and training objectives. *Source*: Adapted from Helen McPhun (2012) Sometimes you get what you deserve! *Learning Matters*, August.

 **ACTION PLANNER**

Revisit Action Sheet 4.1 in the free Action Planner that accompanies this book to add to the list of things that may indicate that training is *not* the solution to your environmental problem – or, at least, not yet.

Organizational culture: the iceberg effect

Company culture is like public relations (PR): it's something you have whether you plan it or not. And, for many organizations, it's unspoken and implicit, as I've seen in my work on manufacturing and construction sites: on one site, people will walk past a faulty environmental control, while staff on another will call the environmental team or simply fix it themselves. How much harder will it be for an individual to break with the group dynamic in the former case and fix or point out the fault in time to prevent a polluting spill?

Ed Schein depicts organizational culture as an iceberg: artefacts such as buildings and logos are at the very tip and the rest of the small part above the waterline is made up of the organization's espoused values – the things organizations say in their environmental policies, for example, or on their websites and in promotional materials.[18]

But the 80 per cent below the waterline is like the unconscious mind: here lurk the organization's basic underlying assumptions or values that may only be guessed at by observing people's everyday actions – the things that send the unspoken signal 'This is how things are done round here'.

So the likelihood of environmental incidents reflects the environmental culture of the organization, because it influences what individual employees actually do. I've been astounded by the influence of such apparently subtle things as culture and leadership – and have observed companies turn themselves around simply by deciding they have to improve their environmental performance (see Chapter 2).

> **Case study: maintaining a good culture**
>
> A colleague told me a great story about an actively managed corporate culture. A new employee was recruited from another firm, and it turned out he'd come from a culture of accepting lower environmental performance and was not following his new employer's protocols. He was quietly sent on some appropriate training and the problem was promptly and effectively solved.
>
> Workplace practice is strongly influenced by both explicit and tacit expressions of organizational culture, and can be positively influenced, as it was in this case.

A strong environmental culture means that everyone at every level of the organization observes the values and priorities placed on all aspects of the environment, which aligns everyone's actions towards a common goal. Conversely, if the day-to-day behaviour of directors, managers, supervisors and staff does not support good practice, the environmental culture will be weak even where the company has good policy and structural and procedural controls.

Issues such as these – all very real in busy workplaces – mean that training cannot, on its own, bridge all performance gaps. This is the case regardless of whether you are an internal trainer working for the same organization as your trainees or an external trainer with little or no influence over the workplace(s) to which your trainees go back.

You therefore need to use your compliance monitoring systems and regular training programme evaluation to look for the presence and seriousness of any factors that promote or impede the on-site uptake of new practices. There are some important cut-off points to observe between things we and our trainees can control and things we can't. These affect the effectiveness of our training. If there are significant barriers to uptake, you need to find out what they are before going too far down the path of scoping your training programme. This will help you, the individual workplaces and the wider industry to resolve these barriers.

But how to address such barriers to the effective uptake of your wonderful training? It all comes back to *partnership*. As an environmental agency, if the wider industry you're working with understands the importance of better environmental performance in legal, business and environmental terms, its members will do their best to support the application of new performance standards on site. Even as an internal trainer within a commercial or other non-government organization, you will also need to work with other key players within and beyond your organization.

In sum, training will be part of the solution for your environmental problem when you win support for the initiative and its new practices and desired outcomes from the relevant internal and external players, thereby ensuring good workplace support for your trainees.

Effective solutions – training or other – address the causes, not just the symptoms, of poor performance.

 KEY OUTPUTS

At the end of this stage, you will have produced:

▶ a training plan that addresses who, what, when, where, why, how and how much; and/or

▶ a management plan to address issues that training alone can't solve.

5.4 ADDIE: DESIGN IN LINE WITH THE BIG PICTURE

> The two most important activities in instructional design are identifying the correct performance problem and underlying cause in the first place, and effectively evaluating your solution. Do these well and the actual instructional design will take care of itself.
>
> Marc Rosenberg[19]

The big picture of designing your environmental training to meet your training needs has two leading elements: the principles and practices of good training; and the characteristics of your trainees and the workplaces where they apply their new practices.

 LEADING QUESTIONS

At the start of this stage, you are asking the following questions.

- ▶ What key aspects of adult learning apply to your audience?
- ▶ How, when and where will they want to learn?

Good design translates into more effective training content and implementation of your training. It usually means less content and more hands-on practice – but it will be worth it if you want your trainees to retain their learning. And you do, don't you?

I must confess that this is the one thing I still struggle with: I love my subject so much that I get very enthusiastic and often cram too much into my workshops. But, most times, less is more! Training is one of the hardest things in the world to do well and one of the most rewarding when you succeed. When you can distil down the key learnings and deliver your training using the principles and practices that the best trainers use, then you will have created memorable training that your trainees can apply with confidence.

Principles of good training

There are many sets of principles that support good learning outcomes. As well as ADDIE, prominent among those favoured by professional trainers and learning and development professionals are the following models.

- **Kolb's learning cycle** Illustrated in Figure 5.3, Kolb's learning cycle encourages trainers to build opportunities for concrete experience, reflective observation, abstract conceptualization and active experimentation into both their training and the workplace support needed to help trainees to practise applying their new learnings.[20]

- **The 70:20:10 model** This acknowledges that around 70 per cent of what people learn is gathered from informal experiences in the workplace and enables trainers to make the most of this.

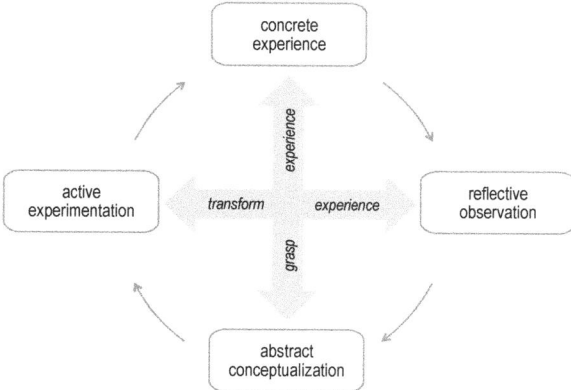

Figure 5.3 Kolb's experiential learning cycle. *Source*: After Alice Y. Kolb and David A. Kolb (2017) *The experiential educator: principles and practices of experiential learning.* EBLS Press, p. 32.

The importance of Kolb's approach lies in its emphasis on repeated cycles of action, reflection and experimentation – which means giving learners permission to fail, reflect, learn and try again, both during the course of the training and when back at work. As David Kolb says, the importance of cyclic, rather than linear (trainer to learner), learning is that 'learners receive information through concrete experiences and transform it through reflection and conceptualization and then transform it again by acting to change the world ... They are both receivers and creators of information'.[21] Only in very supportive workplaces is this possible – and, in the environmental context, safeguards need to be in place to ensure that an incident does not result.

Yes, you can deliver training that is less than excellent – but why spend money doing it? The total losses to businesses from ineffective training have been estimated at a staggering US$13.5 million per year per 1,000 employees.[22] We owe it to our trainees and the environment to invest our training spend as wisely as we can.

 ACTION PLANNER

Use Action Sheet 5.2 in the free Action Planner that accompanies this book to consider how you could build Kolb's learning cycle of into how you design, develop, deliver and evaluate your training.

What makes for great training?

When you understand how to train so that people learn and when you understand your trainees, you will be able to identify the best vehicles to use to deliver your training and how to build in a range of methods to fully engage your trainees in their learning. The academic 'chalk and talk' (also known as 'death by PowerPoint') approach suits very few people – even academics!

In a way, we're all experts at this: haven't we all come out of a workshop totally inspired, full of confidence, and raring to get out there and do it? And, sadly, most of us have also come out of workshops bored, tired and dispirited. When it comes to training, we all know exactly what we do and don't like.

 ACTION PLANNER

Use Action Sheet 5.3 in the free Action Planner that accompanies this book to capture your experience of great and not-so-great training, to inspire you to make your training the best it can be for your trainees. Then come back and use your work to interpret and critique the summary that follows.

For schoolchildren – and I can't see why adult learners would be much different – the key to effective teaching is the quality of the feedback they get on their learning and their interaction with teachers. John Hattie's meta-analysis ranked 138 aspects of schooling derived from 50,000 previous studies of 83 million students from around the world.[23] He found that student–teacher interaction at schools overwhelmingly emerged as the top indicator of teaching effectiveness.

To paraphrase John's finding for our purposes, the leading indicator of effective training is 'self-reporting' on learning – that is, when the trainee knows exactly how well they are doing and can explain this, together with any gaps in their understanding, to the trainer. It's when this happens that a good trainer can give the most useful feedback.

The following are some good practice principles for schools from Mark Nichols that I've adapted to fit adult vocational training – and you can see how John Hattie's findings fit in with them.[24] Aim to:

- communicate high expectations and clear learning goals and the path towards them;

- provide for and respect diverse talents and ways of learning;

- tailor the balance between challenge and support to trainees' readiness and potential;

- create an environment that supports inquiry by encouraging trainees to engage with the trainer;

- encourage active and cooperative learning among trainees – allow them to spend more time on task than watching or listening to the trainer;

- emphasize time on task by using deliberative practice – that is, allow trainees time to carry out difficult and repeatable activities with a well-defined end goal and give prompt, supportive and informative feedback on the quality of each repetition;[25]

- elicit active and critical reflection by trainees on their growing experience base;

- link trainees' questions to genuine problems or issues of high interest to them, to enhance their motivation and accelerate their learning; and

- take every opportunity to develop learners' effectiveness as learners – praise them for asking questions, supporting each other, offering ideas and practising tasks.

These principles and practices will give your training a strong pedagogical basis. For my own face-to-face workshops, this means keeping them small, with ideally no more than 20 people attending. I ask the trainees to divide themselves into groups in such a way that each group comprises a mix of professions or members of different organisations. Each group carries out several practical activities over the one- or two-day workshop and reports on its findings to the other groups. This ensures that people learn from their own and each other's experiences.

Understanding your trainees: creating an avatar

There is a risk that we deliver training to our target audience in the way *we* like to learn – but not the way *they* like to learn. The better you can characterize your trainees, the better your training will be. To help us to understand how they will want to learn, we need to know who they are as people and what support they need to make best use of the training.

The first thing to do is ask your potential trainees themselves what will best suit them. For example, a week-long intensive may be the most efficient delivery mechanism for a trainer and their project sponsor, but many organizations can't spare key people at trades or professional level for such a long time. If you can't access your trainees directly, ask your partners how a given target audience will like to learn. You will find that their questions will drill down to a surprising level of detail about exactly who needs to attend your training.

 ACTION PLANNER

If none of your partners can describe your trainees in detail, then perhaps you don't have all the right partners for your training programme around the table? If so, go back to Action Sheet 3.2 (the partnership worksheet) in the free Action Planner that accompanies this book and update it to ensure you have full representation of the people who are interested in and affected by your training.

Once you have enough detail, you can develop an avatar (some people will be more familiar with the term persona) to help you to better understand your trainees' learning preferences – that is, when and how they will best absorb, question and practise your training. An avatar can be described as a fictitious character who represents one of the many different real people who make up your target audience.[26] A good way of generating an avatar is to base it on a real-life person whom you know, who is one of your target audience. Can you clearly bring this person to mind? Can you see them? Can you hear their voice? You may need to create more than one avatar within each target audience to pin down their respective training needs more tightly.

Avatars can be synthesized from data collected from one-to-one or panel interviews with people who represent your real group(s) of target audiences and I strongly recommend this approach. Capture the information you need in a narrative: this can be anything from a couple of paragraphs to a two-page description. Add a few fictional personal details and perhaps even a name – making sure to keep it respectful, because this will set the tone for your training – to bring your avatars to life as realistic characters who are truly representative of your target audience.

Working with your avatar, you can ask yourself questions about your trainees' workplace, attitudes and actions, their goals, desires and fears, their likes and dislikes, and their limitations and constraints. What are their physical, educational, personal and social characteristics? What is their experience? How willing are they to learn? What sort of learning situation would be best for them?

As you develop your avatars and they take on their own life, you may start to feel uncomfortable as you realize how much you need to tailor your training delivery to meet their particular learning styles. This is a good thing: their strengths may be very different from yours, so if you feel awkward and unsure about trying to come up with a kinaesthetic method rather than a visual/auditory method, for example, it's a sign that you're reaching outside of your comfort zone. Conversely, if you are cruising along with a comfortable formula, ask yourself whether you are really tailoring the training delivery to meet your trainees' likely learning preferences. Some people will learn in the way that you prefer to deliver, but others won't.

The last time I did this, by interviewing a leading hand about his staff, I gathered a whole lot of other vital information that helped me to split the set of training needs I'd identified into two very different groups of people. My client and I avoided wasting our time on training that suited neither, and the two different sets of training that I delivered were consequently more tightly focused on each of the two sets of trainees.

Another top tip is to mix and vary the methods that you use, seeking feedback, so that you can stretch and grow as a trainer. If you do this thoroughly, you will end up with training that is truly user-centred – and a wonderful experience for everyone involved.

 ACTION PLANNER

Use Action Sheet 5.4 in the free Action Planner that accompanies this book to develop your avatars and gain a deeper understanding of how they might like to learn.

Issues that affect our trainees' ability to learn

Astonishingly, around a fifth of adults in almost all countries and economies can't read well enough to participate fully in work and daily life: 18–40 per cent have poor reading skills; 20–50 per cent have poor numeracy skills; and around 25 per cent have no or limited computer skills.[27] Yet environmental work needs to be carried out by people who are competent with both the words and the numbers in their technical guidelines or internal performance standards. On the plus side, employers who provide literacy training reap huge rewards in loyalty and productivity. For one manufacturer, on-site literacy training increased productivity in one department (measured by the optimum time taken to make a given product) from 60 to 90 per cent.[28]

Growing awareness of the ways in which all forms of diversity, including neurodiversity, can add value in the workplace makes it essential that trainers think sensitively about the ways in which diversity can impact on our training methods and delivery. The good news is that making our training more inclusive of diverse learning needs results in better learning outcomes for everyone and improves access not only to training, but also to the environmental and sustainability profession.

There are very real environmental, legal and safety issues on rapidly changing worksites that 'by definition are frantic with non-standard, high risk procedures', which present 'the highest level of risk for mistakes, near misses and ultimately accidents'.[29] In these cases, we can't make any assumptions about people's understanding: to keep them and the environment safe, we need to be sure they understand.

The moral of the story? In some cases, we need to do some research – including through our avatars – to work out the barriers to learning posed by difficulties with language, literacy and numeracy, so we can bridge them before or as part of our training.

In addition, when facilitating the round of introductions at the start of a workshop and canvassing particular objectives that trainees may have from attending, I always make a point of encouraging people to pipe up and tell the rest of us if they are really too busy to be there, or just found out the night before they were going to be there (it's amazing how often that happens), or anything else they need to get off their chest. It always generates a good laugh that opens them up to being there for the day. And, during a tea break, I may approach other people who seem less than fully engaged to chat to them about their work and break the ice, if I can.

Being aware that we are working in a globalized world and that our trainees may have language or other barriers to learning means we can take steps when designing our training delivery to ensure that it's accessible and inclusive from the outset.

Case study: overcoming language barriers on a construction site

This case study is based on part of an excellent conference presentation by environmental manager Amanda Davies.[30] Amanda's paper interested me because it tackled a major issue on many worksites all round the world that doesn't get much attention in the technical literature, but which poses a significant risk to the effectiveness of environmental programmes: the language barriers we may face in a world of very mobile workforces. We have much to learn from Amanda's experience.

The site and the project

Amanda's company was undertaking an underground cable installation as part of a major strategic electricity grid upgrade. The project involved trenches 2.3m deep and 1.5m wide, with cable joint bays every 700m. In total, 63.3km of cable and 32 joint bays were installed across a mix of environments, ranging from open farmland and rural roads to densely built urban areas and ecologically sensitive streams.

The training problem

The team included a large and rapidly recruited workforce of people who had not undergone the firm's standard environmental training. Amanda's team gave staff detailed work instructions and training to make sure that they understood the purpose of the environmental controls and how to build and maintain them to the specifications set out in the environmental permit.

Nonetheless, time after time, the controls were not built to spec, with lots of minor problems necessitating repair or reinstallation – a wasteful and expensive process. When asked if they understood why it was important to get the controls right and whether they were quite sure they knew how to do so, the staff always assured her that they all did – yet the errors were frequently repeated. Why?

It turned out that most of the new staff actually doing the construction work were from Tonga, a group of tiny Pacific Islands nearly 2,000km north-east of New Zealand. While fully numerate and literate in their own language, English was their second language and the technical terms used to describe what they had to do on the construction site were beyond their English vocabulary.

There are many reasons why willing students say they understand when in fact they don't, including cultural factors, a desire to please, a desire to avoid embarrassment and so on. While a good trainer will ask trainees to explain their understanding of the instructions or other learning content, we need to be sensitive to the possibility of barriers to understanding such as second languages, and low levels of literacy and numeracy.

The training solution

As soon as Amanda and the team realized the cause of the performance gap, they put in place a highly practical training programme with lots of hands-on demonstrations. The site environmental team worked side by side with the construction crews on site, teaching them best practice techniques for installing and maintaining the environmental controls. They translated written material into Tongan and made intensive use of visual aids to communicate best practice on things such as the company's ten mandatory safety rules. Staff were taken through these rules verbally and then handed a printed pocket card to carry with them as a reminder of the required personal safety behaviours.

Toolbox talks are informal daily on-site training sessions. Amanda's team made sure that environmental issues were given good airtime and the translation made these discussions accessible to the Tongan staff. Within a matter of weeks, their performance had dramatically improved.

What resources were needed

This change demanded a commitment of time and money. It took several weeks to identify the language barrier, but it was bridged within only a few weeks. Amanda says that the return on investment (ROI) was more than worth it: as well as significantly reducing environmental risks, the positive results of investing in staff development paid off over time by fostering more skilled and committed staff.

Designing your training delivery

In the design stage of the ADDIE training model, you will also consider how to deliver (implement) your training, for example via:

- face-to-face engagement in 'classroom'-style interactive workshops;
- field trips to a real or hypothetical worksite;
- demonstrations on site or in workshops, such as of flocculation in action;
- the trainees' own workplace(s); and
- electronic tools (known as e-learning).

The first two vehicles are classical 'off-the-job' training that takes place outside of the normal workplace. This allows trainees to get away from work to a place where they can concentrate more thoroughly on the training itself and meet new people who may do things differently in their own workplaces. This type of training is thought to be more effective in inculcating concepts and ideas, with practical activities in the field adding a hands-on dimension.

Remember to meet all health and safety requirements. In 25 years of training, I've experienced only two on-site incidents: one in which the manager was alerted to a large swarm of bees heading off to a new nesting site and hustled us all into the office until the bees had safely passed overhead; and a second in which a trainee had to be driven back to the site office because of an asthma attack and prudently had with her the medication she needed.

Workplace training can involve formal and/or informal on-the-job training.

- Formal on-the-job training is often associated with achieving a qualification and usually involves the trainee carrying out normal workplace tasks, with the usual equipment, systems and documents, with the help of the training materials of the technical training provider. Such training is generally regarded as most effective for vocational work.[31]

- Informal on-the-job training, for things such as environment and safety, is often done in toolbox (early morning pre-start) meetings and can be very effective at raising awareness and skill levels.

E-learning – or 'pedagogy empowered by technology', as Mark Nichols calls it[32] – is naturally suited to distance and flexible learning. In the first flush of technological enthusiasm, some people thought that e-learning could wholly replace all other forms and be much more effective. The recent proliferation of massive open online courses (MOOCs) lends some credence to this, with more than 40 top-line US universities and others worldwide, as well as a growing number of private institutions such as the Khan Academy, providing free or paid online learning.[33]

However, much environmental training involves hands-on work, meaning that an on-site component is essential. This means that a mix of training methods, known as 'blended delivery', will support enjoyable, accessible and effective learning that directly relates to trainees' workplaces.[34]

Rethinking training design: the advent of micro-credentials

Training providers are having to do a major rethink of their delivery. The capacity of organizations to release their staff for in-house or external training is much more curtailed than it used to be: people are reluctant to spend two whole days, or even more, away from work. It's hard to get people to attend even a one-day workshop. Much training is being reformatting into half-day or one- or two-hour sessions – or even 10-minute sessions.

Digital micro-credentials are emerging to enable busy people to acquire proof that they have built up particular skills that haven't been formally recognized. Often sponsored by recognized learning institutions, each assessment is small enough to be manageable for busy people, but big enough to be meaningful to employers.

Interestingly, these 10–60-minute modular training formats help us to identify a sequential series of learning steps that often build on some basic learnings that are common to several different training topics. Once those core topics are covered, a modular series of shorter training events can be prepared from which trainees and/or organizations can select those that best meet their immediate needs. Such incremental steps reinforce and build on prior knowledge without tiresomely repeating important, but basic, knowledge.

It's all another reminder of the critical importance of focusing on trainees' and their organization's business needs and making the case for why environmental and sustainability training must meet personal, professional and organizational needs.

E-learning is still 90 per cent learning!

It's important to appreciate from the outset that e-learning is not a quick fix and that, while it may be cost-effective, it is not necessarily cheap. More importantly, the core tenets of education do not change when e-learning is applied and e-learning practitioners must be careful to base their practice on identifiable learning theories – that is, on good pedagogy. Rather than 'trying to replace theories of education, e-learning creates new possibilities for applying established educational and interpersonal theories'.[35]

E-learning is text-light, image-heavy and interactive, so it's great for people with literacy, numeracy or language barriers to learning – all very common in today's workplaces. It's also particularly good for very practical applications where images are better than text – often the case in environmental training.

Case study: I can't use e-learning because my trainees can't read!

Interestingly, language and literacy problems are not a barrier to e-learning; with good management, neither are digital literacy barriers or ability to access the Internet.

Well-designed e-learning can be very effective at transcending language and literacy issues. Research has shown that e-learning, when blended with face-to-face learning, is effective for literacy and language training for diverse Pasifika peoples and other non-English-speaking people living in New Zealand.[36]

Employers may think e-learning is good because staff can do their learning at home – but we all have such busy lives that this won't necessarily work well. On one retail-based e-learning project, everyone agreed that workplace training is legitimately done at work, so a computer was put at stand-up height in every store. Staff were able to access their training during slow times at work and when they needed to ask a particular question to help a customer. No problems with home Internet access there – and a more relevant learning experience in the workplace.[37]

E-learning can be cost-effective when:

- large numbers of people are involved;

- the training content is highly consistent across the target audience;

- working people want to study in their own time;

- people learning at work can pause their training without losing their place to attend to tasks or customers;

- you want to give your trainees regular positive feedback and track their progress through and success with the training materials; and

- large distances and dispersed workplaces make it hard for trainees to get together and too expensive to get trainers to them.

If you're thinking of exploring this option, you need to consider three things that involve three very different kinds of people, as follows.

- **Subject**   This is the province of the 'subject matter expert' (SME) – that is, the person who deeply understands the environmental topic, although they may not be an experienced trainer or familiar with e-learning methods.

- **Pedagogy**   This is the province of the training expert, who has a deep theoretical and applied understanding of how people learn and consequently of what constitutes effective training. Only after we're clear on these first two things can we look at what electronic delivery methods will do them justice.

- **Delivery platform**   This is the province of the developer, the specialist information technology expert who will set the training up in an electronic learning management system (LMS) – the software that sets up how the training material is presented, the level and type of interactivity for the trainees, and the degree to which the trainers can monitor the trainees' work.

In this last regard, remember that bad training will still be bad training regardless of how many bells and whistles are plugged in. The quality of the training reflects the quality of the pedagogy, regardless of the platform. Make sure that your e-learning experts understand good pedagogy as well as good technology.

Remember too that workforces and workplaces are changing, with multi-generational teams working in different sites (not all of them offices) and in different countries.[38] Generations (Gen) X, Y (more commonly known as millenials) and Z are particularly tech-savvy and independent, and want meaningful work. Raised with electronic devices, millennials expect and demand their use, when and where they want them, as part of their vocational training. Many businesses and learning institutions have embraced this approach and ever more sophisticated social media reinforce the need to understand that most of our learning is experiential, not theoretical. Informal, less-structured communication is becoming increasingly important, and many training programmes now supplement delivery with internal social networking technologies.[39]

Not only is learning becoming both more social and more virtual as technologies enable this, but also it's becoming much more collaborative – not only among learners, but between learners and trainers – as learners take responsibility for their own learning and ask correspondingly more of each other and their trainers.

E- and m-portfolios are increasingly being used as part of training to demonstrate on-the-job skills as part of gaining a qualification. Consider this scenario.[40]

A worker on a big construction site is doing his routine inspections of the environmental controls. He has an electronic checklist based on a geographic information system (GIS) that he can tick off and comment on, so that he can document and report on what he finds. He also has a smartphone with a camera and the ability to access the diagrams for every control measure.

1. He sees something wrong with a large control measure and dials up the guideline to check what it should look like, to compare it with the as-built diagram.

2. He takes a photo of the fault and uploads the image to an online register, which records the measure's name and location and the time and date of the photo.

3. The upload alerts the environmental team that a repair is needed.

4. The next device is smaller, but also has a minor fault, less serious than the other.

5. This time, he photographs and uploads an image of the fault, then fixes it and uploads the image of the repaired and now fully functioning device.

6. He also uploads the last two photos to his e-learning portfolio as part of his work towards a work-related qualification.

How cool is that?

> **Case study: capture the learning electronically and use it for training**
>
> After four failed attempts to build a grassed swale along a motorway to the design and performance specifications, environmental manager Kylie Eltham told me she finally got the project engineer to agree that maybe the topsoil had to be sieved after all and that perhaps it was indeed a good idea to mark out the dimensions for the digger driver to follow.
>
> Roadside swales are notoriously difficult to build for long-term operation and trouble-free maintenance. On her fifth try, Kylie got what she wanted: a swale that was properly shaped, compacted, topsoiled and planted, and which didn't come apart in the rain.
>
> I suggested to Kylie that, next time the now-expert team built a swale, she could stand nearby, with a colleague filming everything on video, while she explained exactly what they were doing, how they were doing it and why it had to be done that way (remembering that noisy machinery may demand that they do a voiceover later on). Even better, she might get the operators and other people actually doing the job to explain how they do their work. Then I suggested uploading this video to the company's intranet as part of their training and quality procedures, so that this knowledge could be easily passed on to others.

Blended delivery of environmental training has tremendous potential, especially for trades professions within which large numbers of people – well beyond the capacity of environmental agencies and even the largest of companies to deliver – need standardized training.

 ACTION PLANNER

Use Action Sheet 5.5 in the free Action Planner that accompanies this book to think about the mix of delivery tools you can use in the short, medium and longer terms to create a mix of enjoyable and effective training experiences for your trainees, and a list of criteria for assessing and comparing the relative costs of each tool.

70:20:10: optimizing informal learning at work

When I was delivering public environmental training workshops, it always used to worry me that we might be sending our trainees back to an unsupportive workplace that was beyond our influence. If managers and supervisors send their staff off to do environmental training with a negative comment and disparage their efforts to apply learning afterwards, it makes our job of changing workplace practices very difficult indeed.

Long-time professional trainer Colin Dawson puts it even more strongly:

> It is irresponsible for anyone in our profession to design, develop and deliver a learning solution that fails to take into account the support infrastructure needed for learners to perform successfully in their work whenever and wherever they are called upon to do so. It isn't acceptable to simply throw learners out the classroom door into the workforce and then just hope that what we did during the event is enough. It's not. We know full well that learning doesn't stick unless you make provision to support performance in the workplace.[41]

Colin says that the good news is that influencing workplace performance support doesn't take more effort than we're already investing in our training programmes. What it does take, he says, is a shift in mind-set that reminds trainers to set up performance support for their trainees.

Even if they are supportive, middle managers and supervisors will likely need tips and templates to help them to coach your trainees to apply their new learnings and to help you to evaluate the success of your training. This re-emphasizes the need to engage early on with and gain the support of all the key players in the sector for your training, because they need to appreciate that they share responsibility for the effectiveness of your training.

Here, I'll cover how we can enlist some organizational support for your trainees before they come to your training and when they get back to work, even if they work for different organizations. This involves the concepts of formal and informal learning.

What's now called the 70:20:10 rule arose when a number of researchers in the 1970s–1990s converged upon an understanding that 'the odds' were that professional learning and development takes place roughly like this:

- **70 per cent** informal, on-the-job, experience-based, stretch projects and practice;

- **20 per cent** coaching, mentoring and developing through others; and

- **10 per cent** formal learning interventions and structured courses.[42]

In other words, 90 per cent of professional learning and development takes place not in our wonderful training, but at work and mostly in unstructured casual ways. Yet too many trainers still invest most of their time, energy and dollars in inverse proportion to the 70:20:10 rule.

In the past, informal learning was seen as beyond trainers' control, but this is increasingly changing for two main reasons:

- some training outcomes are too important to leave to chance; and

- increasingly sophisticated tools are emerging for influencing informal learning.

Online compliance trainer Tim Hird says that 'compliance training is too important to leave to formal learning alone, particularly given the struggle organisations have in engaging staff with the traditional methods'.[43] He thinks that the 'just-in-time' relevance and higher knowledge retention of social (informal) learning gives it a huge role to play in 'fostering and embedding a culture of compliance that … regulators are looking for'.[44]

Simply becoming aware of the 70:20:10 rule means that we can consciously shift the ratio by making the most of informal learning in a 'formal' way.[45]

And the reason why we might do so is that companies whose managers take responsibility for mentoring and managing their staff and who create the conditions for their staff to excel find that the resulting higher employee engagement generates 'on average 9 per cent more profit per employee and double the revenue growth of other organisations'.[46]

All this helps to build a learning organization that supports and encourages inquiry – just what we need in a fast-changing world.

 ACTION PLANNER

Use and adapt Action Sheet 5.6 in the free Action Planner that accompanies this book to help those supervising your trainees to rehearse and review their new learnings back at work in a safe learning environment.

Case study: building in workplace support before delivering the training

One of my clients identified the need for digger drivers to be trained in how to install particular erosion and sediment control devices that were consistently scoring badly when inspected by auditors. But, before training the digger drivers, we decided to train the:

- designers and their supervisors, in case they were misinterpreting the guideline;
- project manager to ensure enough time is allocated for the installation;
- site supervisor, who would be instructing the digger driver to install the device to the design specifications, on how to give them some positive coaching – and how to evaluate the effectiveness of our training (more on which in Chapter 6); and
- internal environmental auditor, who inspects and scores the quality of the design, installation and maintenance of the device.

We also had to make sure that the senior management team and board of directors fully endorsed the training. Simply training the digger driver alone without ensuring support all along that line of responsibilities would make it much tougher for them to cost-effectively apply their new learnings and hence we trained this very important person last, to make sure that support for their improved performance was in place at the outset.

Think about it: what percentage of all the essential activity outlined in the case study would you define as actual training and how much as other aspects of the 70:20:10 model?

For public training, where you and your trainees don't work for the same organization, you can adapt these processes by sending coaching support information to people who register themselves or others for your training and encouraging them to use it. And you can both take advantage of the informal learning that happens before and after your formal training if you send pre- and post-workshop activities to them.

The 'flipped classroom' model shows how a simple change – sending key information before the training – vastly improves learning outcomes.

Case study: 'Oh no, my teacher's on YouTube!'

New Zealand science teacher Chris Clay won a global award for his innovative teaching when he put podcasts of some of his high school biology lessons on YouTube. His students were initially embarrassed about it all, but they soon adapted and their learning outcomes rapidly improved.

Clay also created a wiki 'so his students could collaborate with one another about what they got – and didn't get – about the podcast. Often students would answer questions for each other'.[47]

This meant that the roles of lessons and homework were effectively reversed:

- homework 'was to watch a podcast and post to the wiki; and
- class time was to "contextualize" the learning'.

As well as making the best possible use of the face-to-face time in the classroom, other benefits of this blended delivery model included:

- engaging students in the online world – one they were familiar with;
- providing personalized learning – students could watch the podcasts in their own time and at their own pace;
- offering a collaborative environment – the wiki – for students and teachers to clarify understandings;
- building relationships;
- improved student achievement; and
- benefits accruing to Clay's own teaching practice from feedback given on his online resources.

Could this 'flipped classroom' approach help your trainees?

 ACTION PLANNER

Use Action Sheet 5.7 in the free Action Planner that accompanies this book together with the information at the start of Chapter 5 to consider what preparatory activities and on-the-job support your trainees need to make best use of your training and how you can exert a positive influence on their workplace support.

 KEY OUTPUTS

At the end of this stage, you will have produced:

▶ an overview of the pedagogical principles informing the training; and
▶ a clear design template or concept map of how it will be structured, delivered and evaluated, possibly including a blueprint, storyboard or prototype.

5.5 ADDIE: DEVELOPING YOUR TRAINING

> A lot of hard work goes into making something easy.
>
> Alan Marston, Planet Communications[48]

Here, we'll learn more about 'less is more'. Trimming your training to the heart of what people really must know will make it more focused, memorable, enjoyable and relevant – those tough tests by which adult learners will evaluate your training. You can read this part of the book alongside Chapter 6, to make sure that your training is set up in a way that is capable of being evaluated.

LEADING QUESTIONS

At the start of this stage, you are asking who you could invite to test your choice of structure, detailed content, learning outcomes and training delivery modes to make sure that the training suits the needs of your trainees and delivers your required outcomes.

Now that you've done your assessment and analysis, and you've created your avatars and designed your training solution, you can start developing your detailed training content.

Framing learning objectives and outcomes

People want good training, but won't attend until they understand its direct and applied relevance to their work. We need to say very clearly *why* they should attend training.

Each course and each of its component training segments, be they a 10-minute micro-credential or a 60-minute module in a one-day workshop, should set out what the trainees should be able to do afterwards, just as defined in your training needs assessment.

This is not as easy as it sounds.[49] The better you can define your learning outcomes in the context of the real-world issues that your trainees face at work, the more enjoyable, relevant and effective your training will be.

Worried about the difference between objectives and outcomes? I was, until I found the following definitions:

- **objectives** are statements of your outputs – what you are setting out to convey to the trainees ('we will tell you X'); and

- **outcomes** are statements of what trainees will know or be able to do after the training ('… and then you will be able to do Y'), expressed as what you will evaluate.[50]

The structure of a learning outcome is usually expressed as:

'By the end of the workshop/module, you will understand [insert knowledge] and be able to [insert verb].'

But what technical content should you include in your training? While I firmly believe that most experts are naturally good trainers, sometimes it can be hard for an expert to identify the very small population of things that trainees must know from within the vast population of things the expert knows – and just loves talking about.

Give them too much information and your trainees will experience cognitive overload. Your learning will be more effective if you make it very clear to them what the few most important things are. If your training is successful, they'll pick up on the other stuff later on anyway, as they develop their own expertise.

The late Kevin Lohan spearheaded a movement for trainers to sort information into two categories. Channelling Yoda, he says: 'There is no "nice to know". There's only "need to know" and "everything else".'[51]

Only a few 'need to know' things should make it into our training. This is the hardest part of selecting the topics to cover in your training workshops: as Kevin says, if we pull out the 'nice to know' elements, then we can use that time instead for valuable content and the practical activities that embed learning.

I've lost count of the number of times I've asked myself, 'What can I take out of the workshop?' Now, I relocate some of the 'need to know' content into a planned series of pre- and post-workshop online activities, freeing up valuable face-to-face time during the workshop and on site. I hate losing data, so I keep a 'bin file' of the 'everything else' that I cut out in case it turns out to be useful later on.

Sometimes, though, we experts forget how it feels to be a trainee who knows little or nothing on the topic, as the following case study shows.

Case study: imagine you have to explain it to your mother

'I watched a training course recently that was designed and delivered by an engineer.' These arresting words headed an article by professional trainer Kevin Lohan about the perils of letting an expert loose on a bunch of brand-new employees on their induction training.[52]

Kevin invented a fictitious widget called a 'leading-edge capering contact' to help him to describe how an expert, who obviously loved his subject, kept telling his trainees much more than they needed to know about installing these contacts. The problem was that his trainees were never going to install them and simply didn't need to know the extra information – but they didn't know how important it was (or wasn't) that they remember it all and were just getting increasingly confused.

Finally, one trainee spoke up, asking, 'What's a contact?' Another asked, 'What's a capering contact?' And a third asked, 'What's a leading-edge capering contact?' 'It was beautiful,' said Kevin. The trainees' questions provided a perfect model for leading people through a new concept.

The engineer – like many of us, used to talking with our industry peers, said Kevin – would have been better off imagining that he had to explain it to his mother (as long as she was not an SME in the same field, of course) and setting up his training in that logical way.

Kevin suggested that the chances of a successful training session would be improved by giving the engineer a very clear briefing on the training needs when asking him to deliver that session and asking him to attend a 'train the trainer' workshop to support clearer delivery.

I love this story, because while experts have knowledge and passion, it's too easy for us to go into far more detail than is needed. This is the thing I still find hardest about training – and I need to recite my own personal training mantra all the time: 'Less is more. Less is more.'

It's better that your trainees learn two or three key things well and have time to practise them than that they come away from the training with a dim impression of a vast array of impenetrable facts.

And if you can't come up with a learning outcome for a topic, perhaps it's a topic that should go into Kevin Lohan's 'everything else' bin?

Preparation time – the 8:1 ratio

A rule of thumb for many trainers and one I have found to work quite well is that it will take you at least eight hours to prepare the materials you need for one hour of contact time in a workshop context.

If you are going to use e-learning, it's my experience that your provider will have the expertise in platforms and pedagogy (the hardware plus the software, and the art and of teaching and learning), but will rely on you for the technical content, so you will still need to allow enough time – your own or someone else's – for the technical input. This is very time-consuming, because the most cost-effective use of an e-learning provider's time is after you have perfected your technical content, leaving them to focus on its presentation and interactivity.

The same applies when you are developing environmental qualifications. One joint initiative between a regulator and a group of utilities on how to handle dry-weather wastewater overflows involved a consultant preparing a very detailed methodology in close consultation with these two parties, as well as the industry training provider. While this was by far the most cost-effective way of getting the job done, such costs do need to be budgeted for.

Developing training materials

The nature of the learning materials and resources you give to your trainees will depend on how you deliver your training. For training that consists of face-to-face and field-based work, you're likely to give your trainees some or all of the following.

- **A workbook** This will include the background information, all the visual images used, and blank pages in which the trainee can make notes and prepare for practical activities. Keep it streamlined and focused on the workshop process and include any necessary detail in appendices.

- **A copy of the relevant guideline** These are increasingly published online only, for ease of updating, and a link should be sent to trainees before the workshop, but having some printed and electronic versions for reference on the day is also good practice.

- **Relevant handouts** You may also offer various handouts from your local regulator, such as leaflets, posters and the various forms used to lodge permit applications. Many trainees won't have seen these forms, and they really do need to appreciate their significance as both legal and technical documents.

For online training, which is usually combined with at least some face-to-face or field sessions, you're likely to give them:

- an online workbook in a format that can be printed out;
- passwords to access the electronic material;
- links to other information, including the relevant guideline;
- access to whatever online training and social media are available; and
- details of a helpline in the event that they need further support.

Remember that the best way of learning is doing, so embed as much practical work and active recall as you can, building on the trainees' existing knowledge and skills. This will make sure that you have fewer modules with less 'stuff' in them – all good for homing in on the all-important things your trainees truly 'need to know'.

 ACTION PLANNER

Use Action Sheet 5.8 in the free Action Planner that accompanies this book to start scoping your training content. After you've developed a full list of contents, apply Kevin Lohan's Yoda principle to sort it into 'need to know' and 'everything else' categories. Develop all your training materials – then apply the Yoda principle once more.

Piloting your training

Who can help you to test-drive your training? There is no substitute for having a live pilot run of your whole workshop. Invite your key internal and external partners to attend. Give them evaluation forms to help them to assess the content, timing, activities and delivery to make sure that each module is as good as it can be and that the proportion of time spent on the different modules is in the right balance – that you're spending the most time on the most important things. Make sure you consider the feedback in full – even the bits you don't like or disagree with!

To make time for people to give their feedback, you may not be able to go all the way through all of the practical activities, but you can explain them and seek your partners' thoughts. If you can, see if you can get some of your partners to come along to your first 'live' workshop to check that you've made the right changes and to identify any more they think might be necessary or helpful.

Don't worry if it takes you some time to get it right. I update my training every time I run it, to build in feedback and ideas from the day, to include new developments, to improve a practical activity, to update statistics or to accommodate other changes. This is part of best practice training.

 KEY OUTPUTS

At the end of this stage, you will have produced:

▶ a full set of training materials in printed and/or electronic form; and/or

▶ a pilot training session that has generated helpful feedback from partners and stakeholders.

5.6 ADD/E: IMPLEMENTING YOUR TRAINING

> We can ... build 'learning organizations', organizations where people continually expand their capacity to create the results they truly desire, where new and expansive patterns of thinking are nurtured, where collective aspiration is set free, and where people are continually learning how to learn together.
>
> Dr Peter Senge, MIT Sloan[53]

Implementing your training is about much more than turning up on the day or launching your online courses. Here, I raise some of the more gnarly issues that have emerged over time for my clients and which I now suggest to new clients they might consider up-front.

 LEADING QUESTIONS

At the start of this stage, you are asking the following questions.

▶ How will you track your training? Using a spreadsheet? A customer relationship management (CRM) system? An LMS?

▶ How will you review each workshop or electronic training module to see how effective your training has been and how you can review your design, development and delivery to keep doing better?

Choosing your trainers

Different training providers suit different situations. You may have found suitable training through a university, technical college, not-for-profit or commercial provider. If not, you will need to draw from the pool of SMEs across the sector – and that may mean that you deliver the training yourself.

Environmental training is usually very targeted and specialized, and despite the growth in environmental qualifications, people with the skills to deliver it have usually built up their expertise on the job over years of hands-on work and, if they are lucky, some mentoring from more experienced staff. Traditionally concentrated in government regulatory agencies, these people are sometimes also found in universities and research bodies, but are increasingly found in businesses and in consulting or contracting firms as well.

Consultants often specialize in particular areas, building up an extensive body of expertise either as individual staff members, by working with clients, or as organizations, by hiring people who used to work for regulatory bodies. Many consultants build up long-term working relationships with clients and know their businesses intimately. This level of trust and the associated practical knowledge and personal relationships can make consultants a very good choice for in-house training or for the delivery of training on behalf of a regulatory, industry or professional body.

I have delivered training as a consultant on behalf of environmental regulators for many years. The advantages include having a certain distance from the organization, which can allow trainees to be somewhat more open and frank. But there are also strong arguments against this model, including that consultants cannot speak for the regulator on policy matters and that the primary relationship the trainees must build is with the regulatory body itself. You can outsource expertise, but not engagement!

> **Case study: to be or not to be (present at your regulatory training)?**
> When Brian and I first started delivering the erosion and sediment control training in Auckland, the council staff decided not to attend the workshops at all. They were concerned that trainees would prefer not to have them there, given the increased level of regulation and enforcement that accompanied the start of the programme and the resentment this engendered in some industry circles.
>
> However, the trainees themselves would say, 'Where's the council? I thought this was a council workshop!', and made it clear that they would prefer the staff to be there. We told the council this and council staff began to attend all or most of every workshop. I firmly believe that this laid the foundations for the excellent relations that subsequently evolved between the council and the industry.
>
> You will need to make this decision on a case-by-case basis. And inviting the council to present at in-house training can build bridges too.

Depending on the environmental issues concerned, other sources of expertise can include people from the sector itself, including those from or actively involved with professional associations, universities (especially those with extension arms) and other research agencies. If they can't deliver the whole workshop, then consider inviting them to make cameo appearances in class or on site. This also provides variety, which we all appreciate.

If you are lucky enough to be able to choose from a pool of potential trainers, how do you recruit and select them? What selection criteria should you use? Two recent stormwater-related projects have grappled with these questions and

I'm grateful to the project sponsors for allowing me to summarize some of their findings here.[54]

The project involved adapting an overseas programme – the US-based National Green Infrastructure Certification Program (NGICP) – to meet local needs.[55] Run by the Water Environment Federation's Stormwater Institute, the certification provides the base-level skill set needed if entry-level workers are to properly maintain, construct and inspect green infrastructure (GI). The programme has stringent criteria for choosing people to deliver its training. All NGICP-approved trainers must have two of the following three qualifications:

- a minimum of a Bachelor of Science (BSc) degree in water resource management or environmental science;

- a minimum 35 hours of experience as a trainer for hands-on, adult learning courses; and

- direct experience in stormwater GI construction, inspection or maintenance.

In other countries, engineering qualifications would also be desirable for GI trainers, and other sectors will have their own sets of skills and qualifications that you and your partners and other stakeholders should consider.

To recruit potential trainers, the project team – all ourselves environmental SMEs – put together a list of people we thought would be great trainers, asked others to add to it and finalized the list. We then drafted a letter of invitation explaining the training we needed and inviting these people to express their interest in delivering the training. We also asked them to recommend other potential good trainers and we invited them all to take part in a two-day 'train the trainer' workshop associated with our annual professional conference.

As well as their technical expertise, we selected these people on the basis of their personal and professional attributes. We knew they were enthusiastic learners and sharing practitioners – some were already delivering training and others were people whom we thought would be great at training – and we needed people who had the time to develop and deliver the training or whose employers were willing to release them to do so. Our client also set up a commercial model to remunerate these people for the training they would deliver.

 ACTION PLANNER

Use Action Sheet 5.9 in the free Action Planner that accompanies this book to start longlisting who your trainers could be and what criteria you will use to select them.

Training your trainers

Remember what the professional trainers say: 'A subject matter expert is a smart, experienced person who can show people how to do the job. All they need is some help from you on how to train effectively.'[56] Whoever you choose to deliver your training, make sure they attend a 'train the trainer' workshop. To ensure that they are credible to target audiences and funders, your trainers must be able to demonstrate that they are up to speed with the principles and practices of adult vocational training. A search for 'train the trainer' will find the courses available near you. Such training makes a huge difference to environmental experts' ability to deliver interesting and interactive training.

Encourage your trainers to join the local branch of your country's professional training association to keep their new skills fresh and current: even good material gets tired after a while and there are always new ways to learn more. You can find some of these associations listed in the Further resources at the end of the book.

One of the most trainer-focused environmental training programmes I've ever seen is the US-based NGICP.[57] The programme requires all its trainers to attend several days of interactive training on how to deliver training on the technical content of the certification programme in line with best adult vocational training techniques.

If you engage consultants to deliver the training for you, they will be keeping up their own continuous professional development (CPD) not only in their subject area, but also as a trainer. You will also need to schedule meetings once or twice a year to make sure that you brief them on any new policy, research or monitoring information they need to know if they are to keep targeting the hot topics of the day.

Who will own the intellectual property?

If you decide to engage consultants to prepare and/or deliver your training on your behalf, you will need to consider matters relating to intellectual property rights (IPR).

The model agreed all those years ago when Brian and I were preparing and delivering our training in Auckland was that we would not be paid to prepare or update the training, but would recoup the costs over time via a standard day fee. This was advantageous for the council: it was able to set out the issues it wanted the training to cover and it incurred no up-front preparation cost – and it took us a very long time to pull all the materials together! The council also had certainty about the trainers' fees, which included all the updating we did as a result of technical or policy changes or making the training better. It was advantageous for us in that we were able to use the materials to deliver training in other parts of the country – subject, of course, to the courtesy of letting the council know whenever we did this. Such work always led to great cross-fertilization and was good for the Auckland workshops as well.

However, there are times when it is desirable for the regulatory body to fund and own the training materials, such as when all or much of the material is already in the public domain, which means that there is no new technical information that needs to be generated, or when a large investment is needed, such as for e-learning. Such costs can be defrayed or even turned into an income stream by licensing others to use the materials.

Where experts are paid to develop training, this is straightforward: the funder has the IPR in the material. Where volunteers prepare new material or where people are already providing valuable paid training, agreements will be needed that also consider legacy aspects when people move on from the training.

I have seen some very good and very bad examples of contractual provisions relating to IPR. Consultants often develop new intellectual property while working for clients and it's unrealistic to ask them never to use it for future work. Contracts should ideally allow consultants to use any new materials for work that is not confidential or commercially sensitive provided that they acknowledge the client. I now include the following in my offers of service to clients:

> In doing this work, I will use my extensive prior experience together with the results of my own research and professional development. This will involve giving you access to some of the intellectual property that I have built up over many years. In turn, I may develop techniques and resources on this project that I may use in future. I assure you that I will never use any proprietary, confidential or identifying information in any of my other work, and that I will only identify you as clients with your express permission.

In some cases, agencies want training materials prepared in a form that allows other trainers to deliver the material. In this case, the deliverables would include the trainee workbook and visual images, and also a trainer's manual explaining the instructions to be given for practical activities and other information enabling delivery of the training using that material. This approach has its benefits, but it is important to realize that no two trainers will prepare or present information in the same way: any trainer worth their salt will, sooner or later (if not immediately), start changing things to make the training their own. Be clear that anyone who can't follow the material shouldn't be delivering the training.

While it is important to be clear about IPR, I believe it is also important for consultants and public agencies working together to discuss the potential situations that may arise and to work together with transparency and trust to achieve the client's training objectives.

Free or fee? To charge an attendance fee or not

Publicly funded agencies need to justify their decisions about whether to charge an attendance fee and, if so, how much. I've seen regulators run workshops for free, for fees that approach commercial rates and for anything in between. Some aim to break even on costs, while others see attendance fees as a legitimate income stream to help to fund other parts of the wider programme. A lot of people think that a fee – even a token sum – will encourage people to be more responsible about turning up on the day and take the training more seriously. Others don't have those concerns.

People will pay very large fees for many kinds of training, some of which I would deem to be vague in the extreme, but for some reason consider that environmental training should be cheap or free. The costs that 'the market will bear' for environmental training may therefore be rather low.

That said, I believe that the benefits that accrue to the wider public from good environmental management in the private sector mean that some subsidy of attendance fees can be justified and I've seen ecological economists present conference papers with data to justify it.[58]

If you're not sure about charging, find out what it costs for your target audiences to attend other relevant training. Ask your industry partners or trades and professional associations, many of which run training for their members, to find out what they charge.

Sponsorship – pros and cons

Sponsorship is a good option to explore. Find out whether your organization has a sponsorship policy and, if it doesn't, talk to your manager about how receptive senior managers and elected representatives would be to the concept. If you don't have a policy on sponsorship, you many end up having to prepare a report justifying its use.

I've seen two different sponsorship models run for environmental training provided by environmental regulators, as follows.

- A **cost-neutral model** sees providers of industry-related products and services put up displays of information, give each trainee information to take away and sponsor refreshments at the end of the workshop. Some agencies prefer this to be non-alcoholic: many dangerous worksites are 'dry' – that is, no alcohol is allowed on them at any time, and this often extends to off-site work-related activities such as training. Others consider beer and wine to be acceptable, provided that juice, soft drinks and food are also provided. The level of funding contributed is enough to cover the costs of the refreshments. In this model, you can introduce the sponsors at the start of the workshop and welcome them back at the end of the workshop to mingle with the trainees.

- A **subsidy-based model** similarly sees providers of environmental products and services put up displays of information, give out information and sponsor refreshments. However, they also donate additional funding that contributes towards the cost of running the workshop such that the net cost of attendance for trainees is significantly lowered. In this model, you can introduce the sponsors at the start of the workshop. They stay all day and may be called upon to answer questions about the performance and cost of particular products, and they may make a presentation on a local case study of interest. They then also stay after the workshop to mingle with the trainees over the refreshments.

Sponsorship constitutes a form of endorsement of the sponsor's products and services, and some organizations will be unwilling to do this. However, in my experience, sponsors can become a long-standing and valued partner in delivering the workshops.

Sponsorship may prove difficult in situations in which there are many potential sponsors and/or the regulator may not have had time to assess their products and services sufficiently that they feel comfortable about effectively endorsing them. In such situations, make it clear to your senior management, elected representatives and trainees, as well as sponsors and the trainees, that the involvement of a large number of sponsors does not constitute endorsement of their products.

There is also a fine line between sponsors explaining or training and sponsors promoting their products. Trainees quickly become disenchanted. I've been lucky enough to find this very rare, but it does need clear communication and agreement.

Consider using Edward de Bono's 'six thinking hats' process to weigh up the pros and cons of a potential sponsor.[59] Think of any sensitive or thorny issues with which a sponsor is or could be associated, so that you avoid focusing attention on the sponsor instead of on your training.

 ACTION PLANNER

Use Action Sheet 5.10 in the free Action Planner that accompanies this book to consider the wider organizational and social context you work in and the availability of potential sponsors who could subsidize the workshop. Consider which model might work best and work through the pros and cons of using sponsorship in your training programme.

Some other considerations

What follows is a list of other considerations for you to take into account when implementing your training. Every jurisdiction is different, so make sure you get good advice on these points and note any others that may arise.

- **Insurance**   What kind of professional indemnity, public liability, cancellation or other insurance does your organization normally require its suppliers to have? Also consider health and safety for site visits, for example, and where the split lies between the liability of your organization and that of your trainer or the site host.

- **Moderation**   How will you maintain parity of excellence across your trainers? What measures do you need to put in place to manage their selection, quality, moderation and evaluation? Here, I emphasize that moderation should be set up as a stimulating and enjoyable process of collaborative learning – a process to be embraced.

- **Relevance**   How will you ensure that your trainers are keeping their materials up to date in terms of policy, regulatory and technical changes, as well as training methods and preferences? How can you ensure that your needs assessment, training design and delivery keep up with those developments?

Case study: creating a welcoming learning environment

I've said very little in this 'I' part of the ADDIE model about how you, as a trainer, will deliver your training – especially in a face-to-face workshop setting. This is mainly because I can't do better than urge you to attend 'train the trainer' training. But let me share one 'top tip': set up well before your trainees start arriving.

Always allow yourself plenty of time to set up the room for 'classroom-style' workshops. I found myself saying to a colleague recently that 'facilitation is all about moving furniture', as we shifted tables and moved chairs to create a space suitable for open enquiry. It's not a simple process to get the room right for the atmosphere you want to create for your trainees.

I like to allow at least an hour to get everything set up, especially if there is electronic equipment to get working and workbooks, handouts and other materials to set out. Some trainees will always get there early, and it's great if there is time to welcome everyone personally into a professional space. This also gives us the chance to learn trainees' names and a bit about them as they arrive, allowing us to engage with them in a friendlier way during the workshop.

Playing music in the background can also make it easier for early arrivals to enter a space, and it's a great crowd-management tool throughout the workshop, too.

 ACTION PLANNER

Revisit Action Sheet 3.9 in the free Action Planner that accompanies this book to see if there is anything you can add to your moderation process that might also be an enjoyable part of your trainers' CPD.

 KEY OUTPUTS

At the end of this stage, trainees will have attended your training and you'll have some feedback from them on how relevant, complete, clear and engaging it was. This will help you to iterate back to review the assessment, analysis, design, development and implementation of your training. If you've asked some open questions in your workshop evaluations, then you may have received some great new ideas for other things that will improve the training.

NOTES

1 Dr Brendan Moloney (2012) Training techniques and skills needed in 'new-age' organizations. *Training and Development*, April.

2 Josh Williams (2019) *Are we falling out of love with university?* An interview with Industry Training Federation chief executive Josh Williams broadcast on Radio New Zealand on 26 July. Available at https://www.rnz.co.nz/national/programmes/thepanel/audio/2018705957/are-we-falling-out-of-love-with-university [accessed 26 July 2019].

3 Jost Reischmann (1986) Learning 'en passant': the forgotten dimension. Available at https://works.bepress.com/jost_reischmann/ [accessed 16 September 2019]. You can find out more about Professor Reischmann's views on andragogy at www.andragogy.net/ and www.reischmannfam.de/ [both accessed 16 September 2019].

4 Shirley Caruso (2010) *Malcolm Knowles and the six assumptions underlying andragogy*. Available at www.eadulteducation.org/adult-learning/malcolm-knowles-and-the-six-assumptions-underlying-andragogy/ [accessed 5 June 2019].

5 R. Bogue (2102). *Training search to be your adult learning hero*. Available at www.trainingindustry.com/learning-technologies/articles/training-search-to-be-your-adult-learning-hero.aspx [accessed 5 June 2019].

6 Sir Richard Branson (2013) Richard Branson on how to train your employees. *Entrepreneur*, 23 March. Available at www.entrepreneurmag.co.za/advice/staff/managing-staff/richard-branson-on-how-to-train-your-employees/ [accessed 15 August 2019 but no longer accessible]. The quote is the last two sentences.

7 O. Blaylock (2012) Survey highlights more need for business training. An article about the Colmar Brunton-David Forman Business Training Survey (sourced through an article by Rob O'Neill in the *Sunday Star Times*, 26 August 2012). Sadly, I have not seen this survey of New Zealand workplaces repeated and the report is no longer available online.

8 The ADDIE model was created in 1975 by the Center for Educational Technology at Florida State University for the US Army and has evolved since then to become more iterative, dynamic and user-friendly as a result of countless adaptations over the years by professional trainers.

9 (1) Eóghan Quigley (2018) ADDIE: 5 steps to build effective training programs. Available at www.learnupon.com/blog/addie-5-steps/ [accessed 16 May 2019]. (2) See https://en.wikipedia.org/wiki/ADDIE_Model [accessed 16 May 2019].

10 (1) See https://en.wikipedia.org/wiki/ADDIE_Model [accessed 16 May 2019]. (2) Eóghan Quigley (2018) ADDIE: 5 steps to build effective training programs. Available at www.learnupon.com/blog/addie-5-steps/ [accessed 16 May 2019]. (3) A series of presentations in 2014 on ADDIE hosted by the Auckland Branch of the New Zealand Association of Training and Development and presented by Beryl Oldham, Rustica Lamb, Anna Kingston, Robyn Brown, Jacinta Penn and Helen McPhun.

11 Beryl Oldham (2014) *Analysis of learning requirements: starting with the end in mind*. A presentation to the New Zealand Association of Training and Development on 14 May.

12 See www.changefactory.com.au/service/developing-people/training-needs-analysis/ [accessed 15 August 2019].

13 Janne Pender (2008) Training needs analysis. *The Networker*, Feb–April.

14 Robert F. Mager and Peter Pipe (1997) *Analyzing performance problems, or you really oughta wanna: how to figure out why people aren't doing what they should be, and what to do about it*, 3rd edn. CEP Press, p. 3.

15 Ibid.

16 Helen McPhun (2012) Sometimes you get what you deserve! *Learning Matters*, August.

17 Ibid.

18 I'm grateful to my friend and colleague Dr Lesley Stone for enlightening me about Ed Schein's iceberg model as part of the training she and our late colleague Greg Brown gave me and others on the Target Zero team. We all worked together for three years on a major waste-avoidance/resource-efficiency project and continued this work in different forms for many years. Lesley is now the sustainability adviser for Auckland University. Find out more about Ed Schein at https://leadership.mit.edu/portfolio-item/ed-schien/ [accessed 5 June 2019].

19 An interview with Marc J. Rosenberg about training. Available at www2.gsu.edu/~wwwitr/interviews/rosenberg.htm [accessed 15 August 2019 but no longer accessible].

20 There is an excellent analysis of David Kolb's learning cycles at www.simplypsychology.org/learning-kolb.html; find out more about Kolb's work at www2.le.ac.uk/departments/doctoralcollege/training/eresources/teaching/theories/kolb [both accessed 5 June 2019].

21 Ibid., p. 33.

22 Alex Khurgin (2015) *How to increase employee engagement using microlearning*. A Slideshare presentation from Grovo. Available at www.slideshare.net/GoGrovo/atd-microlearning-2015-final, cited in Karla Gutierrez (2016) 10 statistics on corporate training and what they mean for your company's future. *Sh!ft*, 28 January. Available at www.shiftelearning.com/blog/statistics-on-corporate-training-and-what-they-mean-for-your-companys-future [both accessed 2 June 2019].

23 (1) C. Woulfe (2009) NZ study challenges world on teaching. *Sunday Star Times*, 4 January. (2) J. Hattie (2008) *Visible learning: a synthesis of over 800 meta-analyses relating to achievement*. Routledge.

24 M. Nichols (2008) *E-learning in context*. No. 1 in the E-Primer Series. No longer available online, but see more recent work at www.ako.ac.nz/knowledge-centre/an-online-orientation-to-open-flexible-and-distance-learning/online-course-an-online-orientation-to-open-flexible-and-distance-learning/ [accessed 5 June 2019].

25 G. Colvin (2010) *Talent is overrated: what really separates world-class performers from everybody else*. Portfolio Trade. Summary from Bosco Hoh's blog, available at http://boscoh.com/books/towards-mastery-deliberative-practice-flow-and-personality-traits.html [accessed 9 November 2019].

26 I am indebted to my friend and colleague Christine Heremaia of Good Causes Ltd for this definition, and to colleagues in the National Speakers Association of New Zealand and my mentor Kim Baird of Amazing Business who also uses and recommends this excellent method of understanding your target audience.

27 (1) OECD (2016) *Skills matter: further results from the survey of adult skills*, OECD Skills Studies, OECD. www.oecd-ilibrary.org/education/skills-matter_9789264258051-en [accessed 17 May 2019]. (2) Literary Alliance (2014) *Stepping up to better working lives: workforce literacy in New Zealand*. Available at www.itf.org.nz/sites/default/files/publications/Literary%20Alliance%20brochure%20for%20web.pdf [accessed 17 May 2019].

28 B. Pont and P. Werquin (2000) Literacy in a thousand words. *OECD Observer*, No. 223, October. Available at www.oecdobserver.org/news/fullstory.php/aid/366 [accessed 5 June 2019].

29 C. Poole (2012) The language barrier is not a safety barrier. Originally published on the SafeToWork website, but no longer available online. Many thanks to Hugh Pollock for sending me this article.

30 Amanda Davies (2012) *What's the big deal? It's only a trenching job?!* A paper presented to the IECA/NZIHT conference, Recent and Future Innovations in Erosion and Environmental Control, 22–25 July, Hamilton, New Zealand.

31 Ibid.

32 M. Nichols (2008) *E-learning in context*. No. 1 in the E-Primer Series. No longer available online, but see more recent work at www.ako.ac.nz/knowledge-centre/an-online-orientation-to-open-flexible-and-distance-learning/online-course-an-online-orientation-to-open-flexible-and-distance-learning/ [accessed 5 June 2019].

33 R. Bencini (2013) Educating the future: the end of mediocrity. *The Futurist*, Mar–Apr, pp. 40–45.

34 C. Feeney, T. O'Regan-Byrnes and G. Ridley (2012) *How blended delivery can enhance the outcomes of the Stormwater Unit's industry training courses*. A report prepared by Environment and Business Group, Innovaid Ltd and RidleyDunphy Environmental for the Auckland Council, June. I am very grateful to the Council for permission to use and adapt material from that report.

35 (1) Nigel Young and Rachel Teasdale are both from Kineo Pacific, and their quotes come from an excellent training session on e-learning delivered on 27 February 2013 under the auspices of the New Zealand Association of Training and Development. (2) M. Nichols (2008) *E-learning in context*. No. 1 in the E-Primer Series. No longer available online, but see more recent work at www.ako.ac.nz/knowledge-centre/an-online-orientation-to-open-flexible-and-distance-learning/online-course-an-online-orientation-to-open-flexible-and-distance-learning/ [accessed 5 June 2019].

36 N. Davis and J. Fletcher (2010) *E-learning for adult literacy, language and numeracy: summary of findings*. Part of a series covering research on teaching and learning in literacy, language and numeracy and analyses of international surveys on adult literacy and numeracy. Ministry of Education report. Originally accessed in June 2012, but no longer available online.

37 Nigel Young and Rachel Teasdale, both from Kineo Pacific, in an excellent training session on e-learning delivered on 27 February 2013 under the auspices of the New Zealand Association of Training and Development.

38 Elliot Masie (2012) *7 key trends impacting organisational learning*. A webinar delivered on 17 April.

39 Tim Hird (2012) Leveraging social learning to improve the compliance culture. *Training and Development*, December. Available at www.businessperform.com/articles/training-practice/social-learning-compliance-culture.html [accessed 5 June 2019].

40 I've adapted this scenario from a story recounted in 2005 at a meeting of the NZ Association of Training and Development by the prescient Elizabeth Valentine.

41 Colin Dawson (2012) Performance is everything. *Learning Matters*, August.

42 (1) M.M. Lombardo and R.W. Eichinger (1996) *The career architect development planner*. Lominger, p. iv. (2) K. Kajewski and V. Madsen (2013) *Demystifying 70:20:10: a White Paper*. DeakinPrime. First edition January 2012; reprinted with amendments March 2012 and June 2013. Available at www.deakinco.com/media-centre/article/demystifying-70-20-10 [accessed 20 May 2019].

43 Tim Hird (2012) Leveraging social learning to improve the compliance culture. *Training and Development*, December. Available at www.businessperform.com/articles/training-practice/social-learning-compliance-culture.html [accessed 5 June 2019].

44 Ibid.

45 (1) M.M. Lombardo and R.W. Eichinger (1996) *The career architect development planner*. Lominger, p. iv. (2) Dr Robin Petterd (n.d.) *Planting the seeds for the 70:20:10 learning model*. SproutLabs. Available at www.sproutlabs.com.au/elearning/planting-the-seeds-for-a-702010-learning-model-ebook/ [accessed 20 May 2019].

46 Results of HR consulting and outsourcing firm Aon Hewitt Best Employers study. Cited in Steve Hart (2012) Time to sack the HR manager: people-first culture sees profits soar. *New Zealand Herald*, 28 November. Available at www.nzherald.co.nz/steve-hart/news/article.cfm?a_id=365&objectid=10850413 [accessed 5 June 2019].

47 C. Barton (2011) Hey look, my teacher's on YouTube. *New Zealand Herald*, 26 November. Available at www.nzherald.co.nz/nz/news/article.cfm?c_id=1&objectid=10768801 [accessed 5 June 2019].

48 From a newsletter in the 1990s from ISP Planet Communications: see www.pl.net/ [accessed 15 August 2019].

49 If you are working without the help of a professional trainer, then I warmly recommend this detailed step-by-step guide: Michael Allen, with Richard Sites (2012) *Leaving ADDIE for SAM: an agile model for developing the best learning experiences*. American Society for Training and Development.

50 J.S. Atherton (2011) *Teaching and learning: objectives*. No longer available online.

51 Kevin Lohan (2012) Not nice to know. *Training and Development*, June.

52 Kevin Lohan (2012) Explain it to your mother. *Training and Development*, April.

53 Dr Peter M. Senge (2006) *The fifth discipline: the art & practice of the learning organization.* Currency, Doubleday, p. 3.
54 I am grateful to Water New Zealand and the chair of its Stormwater Committee, James Reddish, and to the Auckland Council Healthy Waters Team, especially Dukessa Blackburn-Huettner, for permission to summarize their thinking here.
55 Find out more about the National Green Infrastructure Certification Program (NGICP) run by the Water Environment Federation's Stormwater Institute at https://ngicp.org/program/ [accessed 22 May 2019].
56 Bill Cushard (n.d) Subject-matter experts and keeping up with the demand for learning. Available at www.mindflash.com/blog/subject-matter-experts-and-keeping-up-with-the-demand-for-learning [accessed 15 August 2019].
57 Find out more about the National Green Infrastructure Certification Program (NGICP) run by the Water Environment Federation's Stormwater Institute at https://ngicp.org/program/ [accessed 22 May 2019].
58 S. Hatfield-Dodds and CSIRO (2003) The catchment care principle: an integrated approach to achieving equity, ecosystem integrity and sustainable development. A paper presented to the Ecological Economics Think Tank, Auckland, November. There will be much more recent information available on various ecological economics websites and in journals.
59 Edward de Bono (1985) *Six thinking hats: an essential approach to business management.* Little, Brown & Co.

6

ADDI*E*: evaluating the effectiveness of your training

> However beautiful the strategy, you should occasionally look at the results.
>
> Source unknown[1]

Allow at least 5–10 per cent of your training budget[2] to fund monitoring and evaluation of the effectiveness of your training to make sure it aligns with that of the wider environmental management programme of which it is a part and links to the local, national and global monitoring and reporting systems that inform our efforts.

Here, I must reiterate that this chapter is a summary for environmental experts. The depth of expertise of training professionals on this topic is breath-taking, but if you use the resources offered here and in Chapter 4, and if you join a training association or arrange some mentoring from a professional trainer, you will be amazed at how much more rigorous, focused and effective your training will be.

My planetary training model in Figure 1.1 is shown again here as Figure 6.1 because, in this chapter, I'll show you how to evaluate the effectiveness of your environmental training and align it with the globally accepted indicators used by businesses, scientists and governments around the world.

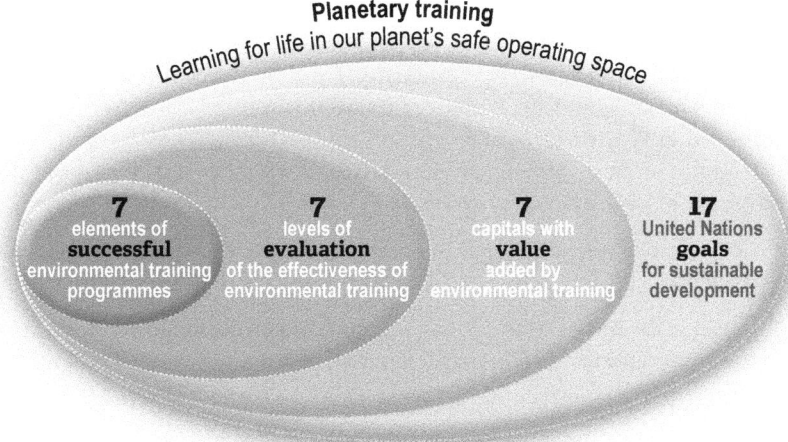

Figure 6.1 Planetary training: revisiting learning for life in our planet's safe operating space

TOOLBOX

Go to www.ESST.institute/Success/Toolbox to download some tools to help you through the ADDIE process. They include a sample evaluation sheet and other standard forms, checklists and templates that will save you time as you set up or enhance your environmental training programme.

 LEADING QUESTIONS

At the start of this stage, you are asking the following questions.

▶ Did the training, as implemented, help us to address the performance gaps, objectives and outcomes we identified in the first two ADDIE stages and our business case?

▶ If not, why not?

▶ If yes, to what extent did we address them and how can we do even better?

▶ Is any further training or other support needed?

6.1 HOW PROFESSIONAL TRAINERS EVALUATE THE EFFECTIVENESS OF THEIR TRAINING

> Remember, training is not what is ultimately important ... performance is.
>
> Marc Rosenberg[3]

Environmental practitioners are good at quantitative monitoring and evaluation. Prepare to be very impressed with the level of rigour that professional trainers bring to this skill.

Professional trainers around the world recognize five levels of evaluation of the effectiveness of training. Levels 1–4 were developed by Donald Kirkpatrick, and Level 5 by Drs Jack and Patti Phillips of the ROI Institute.[4] These five levels are summarized in Figure 6.2.[5]

The work you have done in previous chapters will help you to evaluate the effectiveness of your training programme in terms of:

• how effective your training is in actually changing what people do back at work;

• the contribution of related initiatives, such as new technologies, industry communication and compliance activities; and

• a measured improvement in the environmental outcomes of direct concern, including monetized or otherwise tracked increases in value of this and the other capitals illustrated in Figure 4.2.

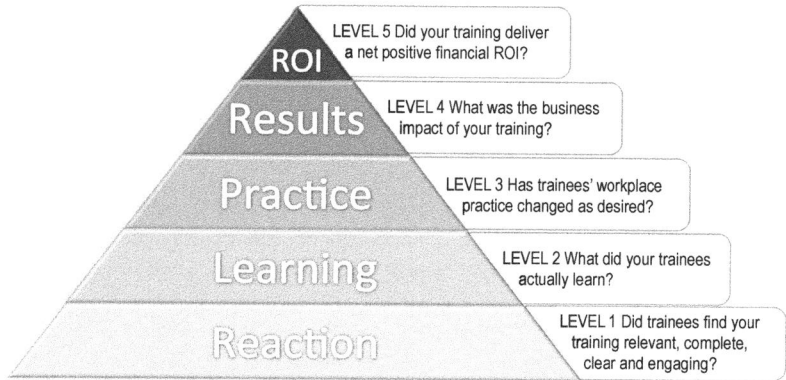

Figure 6.2 How professional trainers evaluate the effectiveness of their training

It's easy to see how the five levels of evaluation can reveal the value added to many of the seven capitals – financial, manufactured, intellectual, social and relationship, indigenous, human and natural – as measured by the Global Reporting Initiative (GRI) and the Natural Capital Protocol.

Remember to ensure that your measures of effectiveness are cost-effective: keep them measurable, meaningful, simple and easy to use, and make sure that they indicate progress towards your identified outcomes. They should generate data that readily allows analysis and interpretation, especially of trends, so as to inform organizational learning and adaptive management.

Case study: how to create a test that everyone will enjoy

According to Evangeline Harris Stefanakis, the word 'assess' comes from the Latin *assidere*, which means 'to sit beside'.[6] Literally then, to assess means to sit beside the learner. This lovely way of looking at learning assessment brought to mind a couple of stories.

Of the many people who've attended my workshops over the years, some have been extremely capable practitioners for whom school would have been one of the least rewarding times of their life. In fact, at one workshop, a delightful young man confided to me that he couldn't read, so please could I not ask him to read an excerpt out loud. It's for this reason that I always call for volunteers to read out loud rather than nominate an individual – and even when someone's workmates are laughing and pointing at one of their team to volunteer, I never accept the offer unless the hand comes out to take the paper. In the case of this young man, I was able to reassure him that I was looking for someone to pretend to be a grumpy judge and that he was far too nice for the job!

Academically inclined people are used to sitting exams and won't be fazed by a written or online test. But for the non-academics who at best are not used to sitting exams or at worst are still scarred by their school experiences, the terms 'test' or 'exam' paralyse them with fear and will set them up to fail.

Several years ago, I did a green building course over a weekend with Johann Bernhardt and Eddie van Uden.[7] They announced that there'd be a test at the end and I was absolutely horrified! Sure, I was interested in green building and wanted to build a green home eventually – but I'm not a builder, so I was convinced I'd make a complete mess of the test.

To my surprise and delight, Johann and Eddie had set up the test to facilitate everyone's success. They handed round a sheet of paper with ten questions on it and space to write the answers (so far, so standard), got us all round a table (a bit of a departure from standard there), and then read out the first question and asked, 'Well, what do you all think?' (By now, this was definitely different from any other test I'd ever sat!) Someone tentatively suggested an answer, someone else asked a question and another offered a story from their experience. At the end, our trainers invited a volunteer to sum up, then we all wrote the answer down.

We went through all the questions in that way, with different people able to summarize the different questions and the trainers making sure that everyone had input. As well as being a thorough evaluation of the workshop, it was also a very enjoyable way of reviewing and consolidating our learnings. It was genuinely a 'sitting beside' experience.

If my avatar work tells me that my trainees will be totally alienated by having 'a test' at the end of the workshop, I flip the concept for them. At the very start of the workshop, I'll make sure that everyone can learn comfortably by saying that, at the end of the workshop, we'll be asking them to tell us how good the training was – that they are testing us, not the other way around. Not only does this get a laugh, but also it's totally true.

At the end of the workshop, I ask the trainees to rate three key things we want to know about their experience of the workshop by drumming their fingertips on the table or whatever we're standing next to and measuring the volume of the drumming on a three-point scale (loud = 3 = fantastic; quiet = 1 = terrible). This is a lot of fun and, together with the group discussion approach, gives very good feedback information.

In this way, I offer choice not compulsion, participation not reiteration and cooperation not isolation – and I model what should be happening in a good workplace.

Who among your trainees and which of your workshops could benefit from this approach?

The following summary of the five levels of evaluation that are of use to professional trainers shows the difference in reach between in-house and external trainers. As an external trainer, I now explain the importance of post-training support in the trainees' own workplaces and help them with that. For agencies training lots of people from different workplaces, I recommend creating a package that is sent to each trainee upon registering for any training, asking them to work through its contents with their supervisor after the training.

Level 1 – Reaction: what did the trainees think of the training?

'Reaction' identifies the extent to which your trainees found the training relevant, clear, complete and engaging. Usually collected via a feedback form (sometimes called a 'smile sheet') or online survey filled out at the end of a workshop, this level of evaluation seeks feedback on these four aspects, as well as workshop delivery, location, interaction, enjoyment and the like. Every trainer and client I know laughs and cringes at comments such as 'The room was too hot' and 'The food was too cold' – but we get what we ask for! Ask the right questions and you'll get very helpful insights into the quality of your needs assessment and training delivery.

Level 2 – Learning: what did they learn from it?

Ask your trainees to list the top three most important things they learned. If this doesn't match what you thought was important, treat it as another piece of vital information to help you to recalibrate your training. You can also test their new knowledge, for example with a short written or online test at the end of the workshop or by asking them to fill out an online knowledge survey within two or three weeks after the workshop. If you are working with people who have language or literacy barriers, use a group feedback session, as in the case study above.

Level 3 – Performance: what can they do as a result of the training?

Performance evaluation is the task of supervisors, managers, human resources (HR) personnel, site inspectors or other auditors who assess what observable changes (as defined in the training needs assessment) they see in trainees' performance as a result of the training. They need support for this, such as check sheets aligned with the training and the guideline, and mentoring on how to have coaching conversations with their staff. With this support, as we saw in the case study 'Building in workplace support before delivering the training' in Chapter 5, supervisors and managers would ideally be coaching and supporting their staff and conducting weekly reviews with them throughout the first month after the training, with the information being collated and sent to the training sponsor and the HR team to help them to evaluate how effective the training was and how it could be improved.

Level 4 – Results: what was the business impact of the training?

Evaluation of the results means asking, 'How much of a change did the training produce for the business? What measurable results have been observed?' This includes measuring the things that you considered in your business case and training needs assessment and analysis, as well as comparing compliance results before and after the training. You could also seek feedback from company staff, external site inspectors and auditors. You may observe environmental efficiencies, site or project efficiencies, reduced staff turnover and a higher rate of winning tenders, clients or customers based on improved environmental performance. For government bodies, you may record less use of expensive enforcement, better water quality results, reduced greenhouse gas emissions and so on.

Level 5 – ROI: did the training produce a financial return on investment?

Return on investment (ROI) analyses, in dollar terms, the benefit that accrues from spending on training and performance improvement programmes. It asks, 'Was there a financially measurable improvement in performance as a result of the training, and was this less than, the same as, or more than the cost of the entire training package?' Here, you use the data you have collected from your Level 1–4 evaluations.

According to Drs Jack and Patti Phillips, the steps involved are as follows.

1. Identify programme benefits (Level 4 business impact).

2. Convert benefits to monetary value.

3. Identify intangible benefits.

4. Tabulate all costs (needs assessment, design/development, delivery/implementation, evaluation and overhead/administrative costs).

5. Calculate the ROI.

Level 5 can include value added to the seven capitals: use the indicator lists from your own policy documents and the GRI. Remember that some improvements, such as reduced staff turnover (where employees stay longer in the job because the training makes their work more rewarding), can take a long time to emerge. For environmental programmes, ROI may also reflect the absence of previous expenses, for example the costs of site shutdowns due to environmental abatement notices or legal expenses in the event of enforcement action.

You won't need to do a full financial ROI on all your training, but the occasional sample will be very revealing. Few professional trainers routinely do ROI evaluation, but environmental professionals are used to rigorous monitoring and evaluation, and my experience is that they cope well with measuring the financial ROI on their training.

Case study: ROI on environmental training – what governments should think about

The cost of my training has got to be less than the cost to the council of taking even one prosecution!

One of my environmental colleagues recently attended my ROI workshop on how to measure the full financial return on investment in environment and sustainability training. She had led a highly successful training programme that resulted in a massively profitable business turnaround for the firm concerned (see the case study 'Training tripled their turnover' in Chapter 2) and was writing a proposal to deliver environmental training for a council.

The council had been experiencing rapid growth and had prosecuted several firms for failing to follow its environmental guidelines. Instead of simply reacting to serious pollution events, the council wanted to set up an environmental training programme to proactively help local companies to protect the much-valued streams and harbour around the city. We talked about the cost of my colleague's work to the council, which needed to allocate the funds in its budget. Suddenly, she exclaimed, 'Hold on – the cost of my training has got to be less than the cost to the council of taking even one prosecution!'

She had realized that if the council were to avoid the costs in staff time and legal expenses of just a single prosecution after rolling out my colleague's training, its ROI would be at least 100 per cent. The cost of preventing two or three prosecutions a year after a one-time delivery of the training amounted to a 200 or 300 per cent ROI per year.

And that was not even counting the ROI that accrued in avoiding the costs of environmental harm to freshwater and marine species and ecosystems, and community concerns about damage to the beautiful natural environment from preventable pollution events or the value of protecting this natural capital.

Back to the realities of the workplace

Regardless of whether you and your trainees work for a government agency, a business or a not-for-profit, environmental training takes place within a wider business and operational context that includes organizational policies and strategies, as well as staff management and support. Because some trainees may find themselves unable to apply their learning when their improved workplace performance is not supported, it is very important to give formal recognition of trainees' learning outcomes by way of a Level 2 training assessment. Trainees who, for external reasons, are unable to apply their learning in the workplace need this recognition – especially in programmes that may deregister people or remove their certification on the basis of workplace performance.[8]

Government agencies delivering training thus need to work closely with their industry partners to support the application of learning in the workplace, for example through site inspection and scoring systems that promote, or even enforce, workplace conditions that deliver the desired environmental performance. Other external trainers and in-house trainers need to encourage the setting up of similar systems to promote the application of learning in the workplace.

All these are opportunities for learning more about how well our trainees are learning. It's a very dynamic and exciting world to be working in!

6.2   A WORLD-CHANGING VISION FOR ENVIRONMENTAL TRAINERS

> The ones who are crazy enough to think they can change the world are the ones that do.
>
> Steve Jobs[9]

Understanding the value that training delivers is part of preparing the business case for funding its delivery. My vision is that environmental trainers will be in the vanguard of the sustainable economy and lead the way in measuring the beneficial outcomes they generate.

The late Alastair Rylatt, a professional trainer based in Australia who gained his PhD on a topic related to ecological sustainability, was a man with the same vision. Alastair identified two more levels of evaluation that professional trainers could use. They are shown in Figure 6.3, and I find them particularly relevant for environment and sustainability trainers.

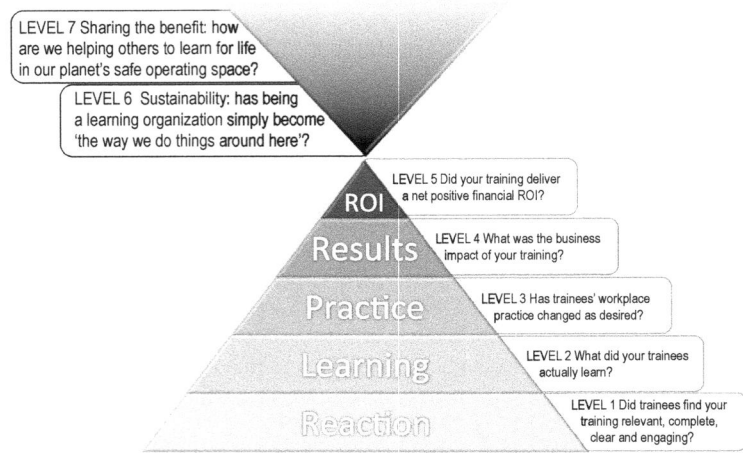

Figure 6.3 Two more levels of effective training

Level 6 – Sustainability: has being a learning organization simply become 'how we do things around here'?

Alastair said that a truly sustainable benefit is realized only when the newly learned capabilities and competencies delivered by training are actually helping to prepare a business for the future. Sustainability also emerges when the effects of the training are fully internalized into a vibrant lifelong and lifewide organizational learning culture, so that the benefits are preserved despite recurring changes in personnel, governance systems and company structure.

In some cases, short-term ROI results can indicate that training did not help the long-term survival of the business. Alastair said that this usually indicates that the training was focused on fixing short-term pain rather than fostering long-term results. He observed that this is especially true when training is not seen as part of a holistic business strategy. Alastair said that addressing business longevity encourages a shift in focus away from short-term expediency, towards issues concerning the lasting social and ecological sustainability of the business.

I totally agree. I've seen companies that embraced sustainability transforming themselves into learning and more profitable enterprises, and such findings are increasingly common in business literature, as evidenced by Michael Porter's and Peter Senge's work and that of a growing number of researchers following in their path. Environmental training programmes are indisputably a catalyst for better business.[10]

Level 7 – Sharing the benefit: how are we helping others to learn for life in our planet's safe operating space?

Alastair Rylatt also proposed a Level 7 evaluation that assesses how we are helping others and the planet through our value chain, the wider society, the environment and future generations. My work in different sectors, backed up by excellent research from all round the world, shows that sustainable firms are up to 24 per cent more profitable than less sustainable firms.[11] So while a few businesses may still cling to the belief that helping the planet is neither their job nor their responsibility, organizations that recognize the part that they must play are invariably more productive and profitable than their industry peers who don't.

> **Case study: turning lives around**
> Graffiti affecting structures, neighbours and machinery on a big state highway project had contractors Fulton Hogan scratching their heads about the escalating cost and disruption of the damage. They hired security guards and put up posters inviting people to contact the city council's 'STOP TAGS' phone line. Locals responded accordingly and several offenders – including a 13-year-old who had caused NZ$60,000 worth of damage – were apprehended and ordered to do community work cleaning up graffiti on the project. The company must have done something

right with those young offenders, because, as site safety manager John Smith explained, 'We turned those kids around ... they have become our eyes and ears because they live alongside.'[12]

The company was also highly commended for taking part in a prison's release-to-work scheme. The 20 prison personnel taken on for the highway job proved 'more reliable than many workers employed through labour hire firms, and six were still working on the project months after leaving jail'. These workers caught buses from prison and Fulton Hogan reported them to prison officers if they turned up late. No one absconded. 'What we found is that they were here at the crack of dawn, every day, regardless. They often beat me to work,' said Fulton Hogan's deputy safety manager, Tash Mullen. 'It gave them that incentive. It gave them something to look forward to. They wanted to be here.'[13]

The US-based Valjean Financing project takes a similar approach: it allows investors to finance, among other things, the re-entry of ex-convicts into training and paid work. Such local support programmes show a better ROI than imprisonment, recidivism and reimprisonment.[14]

These projects make valuable contributions to society in many different ways, across many of the six or seven capitals.

 ACTION PLANNER

Use Action Sheet 6.1 in the free Action Planner that accompanies this book to work out how you will evaluate the effectiveness of your training from Level 1 to Level 7. How high will you aim?

6.3 RECOGNITION OF LEARNING – AND OTHER THINGS

> Creating a growth-mindset in which people can thrive ... involves conveying that the organization values learning and perseverance ... Without [this] belief in human development, many corporate training programs become exercises of limited value. With a belief in such development, such programs give meaning to the term 'human resources' and become a means of tapping enormous potential.
>
> Dr Carol Dweck, Stanford University[15]

Now that you've evaluated your training, you may want to give your trainees some kind of formal recognition of their learning, if you don't already do so. This can encourage people to attend your workshops. Good environmental training is well received by public and private agencies alike, which are increasingly requiring service providers to attend relevant training before they can be eligible to tender for contracts.

Trainees may receive many forms of recognition of their training other than formal qualifications. Some can be given in-house, while others involve working with external parties.[16] Some of the words bandied about include (more or less in order of increasing rigour) 'attendance', 'assessment', 'digital or open badges', 'approval', 'certification', 'accreditation', 'qualification', 'licensing' and 'registration'.

Let's take a look at what these words may mean for your training programme – noting that these words will be used differently in different training and environmental jurisdictions, and that the following summary is intended to give only an overview of some possibilities you could explore. As you go, you'll see how important it is for you to have good support from your HR and information technology (IT) teams.

Supporting staff to gain a formal qualification and celebrating their completion of this and in-house training can open up the world of learning to people who have experienced exclusion from it all their lives. It's one of the most moving and inspirational aspects of my work.

 ACTION PLANNER

Use Action Sheet 6.2 in the free Action Planner that accompanies this book to jot down notes and compare the various options as you read this section on the various levels of recognition of your training.

Attendance

As recently as 2014, a survey of compliance training found that 72 per cent of respondents believed that measuring course completion rates – that is, counting 'cheeks on seats' and issuing an attendance certificate – was the only way of measuring the effectiveness of their training programmes.[17] That's not even Level 1 evaluation!

Over 50 per cent of those same company staff said that they had difficulties measuring the effectiveness of their training – and the authors concluded that this was a weak link in many compliance programmes. And it's widespread: together, workshop attendance and the 'smile sheet' Level 1 evaluation form are still too often the only ways in which many trainers evaluate their training.

Do you need to track attendance? Yes.

Will it tell you how effective your training is? No.

For that, you need to look at the five (or seven) levels of evaluating training we've just reviewed – and possibly some of the following options as well.

Assessment

'Assessment' means measuring each trainee's learning as a result of your training – that is, the Level 2 assessment we've looked at already. Jay Wilson's North Carolina erosion and sediment control training (see Chapter 2) includes a multiple-choice test at the end of each workshop, which people must pass to receive a two-year certification. To pass, trainees must get at least 75 per cent of the test answers right, which 85–90 per cent of people manage first time, while the others can re-sit the test later. This is a basic and extremely useful level of assessment (we'll look at certification itself shortly).

If there were one thing I could change about my past erosion and sediment control training, it would be to create a simple, but robust, learning assessment and to emphasize how important it is for the trainees to complete it. And here's the thing: we did start doing it, but the programme didn't have the administrative systems and capacity to keep it up. That said, with electronic systems and learning management systems (LMSs), it's all much easier now, so I suggest you consider some appropriate forms of learning assessment.

The advent of micro-credentials could also allow assessment and recognition of informal training, such as toolbox talks, which are not formally recognized as training. Properly designed, learning from toolbox talks and informal learning from leaders and peers in the workplace could both be assessed and recognized.

Digital or open badges

Electronic badges can be issued by anyone who delivers training; they don't have to come from a recognized educational institution or certified practitioner.[18] They can recognize people's completion of training and the knowledge and competency gained in what the developers termed an 'electronic backpack', which is visible to the issuing agency and the trainee, and portable for the trainee across changes in employers, email addresses and so on.

Approval

Many trades and professional bodies require their members to complete a specified number of hours or points of approved or endorsed training per year, as part of their continuing professional development (CPD). Depending on the content and duration of the training, these bodies allocate varying numbers of CPD points to approved training courses and members who attend can claim these towards the total required.

Obtaining approval from the relevant organization(s) for people to gain CPD points by attending your training is usually a straightforward and relatively inexpensive process of engaging with the body concerned. They will want to approve your training content and delivery and make sure that it meets their standards for quality and relevance to their members. This is a simple and effective way of making your training more attractive to your trainees.

What organizations could you approach to allocate CPD points to the people attending your training workshops?

Certification and accreditation

Certification is a formal procedure in which an authorized person or agency assesses and verifies that an individual, organization, process, service or product meets certain established criteria, requirements or standards and issues a certificate to say so.[19] The decision for someone to become certified is often a voluntary one. There may be eligibility requirements (such as education or years of experience), an examination and a fee. As we saw, this is the case for professional certifications, such as the US-based National Green Infrastructure Certification Program (NGICP) for green infrastructure. Others include the global Certified Professional in Erosion and Sediment Control (CPESC) and the Certified Environmental Practitioner (CEnvP) scheme, an Australasian initiative.[20] There are usually ongoing requirements, such as regular retesting (say, every two years) and providing evidence of a minimum number of hours' CPD over that time.[21]

Some certifications include the ability to cancel the certificates of those who fail to meet defined professional standards, but this is a demanding process. Back in 1995, the council in Auckland hoped that people attending its erosion and sediment control workshops would submit better environmental permit applications and that their on-site environmental controls would be better, so that their permits would be approved more speedily and their sites would need a lower level of compliance inspections, perhaps by some form of self-certified compliance. This would benefit both the companies and that council. While such improvements have indeed occurred over the years, the council decided not to go ahead with the process of certifying, recertifying and decertifying trainees and/or companies because it was too demanding of resources that just weren't available at the time.[22] However, there is renewed interest at the time of writing and it will be interesting to see what happens next.

There are many levels and types of certification, some of which are as follows.

Certification of trainees   A good example of certification of maintenance contractors comes from North Carolina in the United States.[23] The background was that council staff assessing the maintenance of stormwater treatment devices had found a 90 per cent failure rate and, on investigation, discovered it was because the landscapers contracted to maintain the devices didn't understand their function or what maintenance was needed. When the council required these landscapers to attend a training workshop to gain a certification and to attend a refresher course every three years, the standard of maintenance of these devices significantly improved. This was also a benefit to the landowners because the improved maintenance made for a longer device life – and this equates to a significant cost saving (also a key motivator for development of the NGICP).

Certification of trainers and courses   Certification of environmental training courses is used a great deal in the United States. For example, StormwaterONE is a private company offering a range of online training and certification courses.[24] The training meets the needs of the US National Pollutant Discharge Elimination System (NPDES), and some of the classes are recognized as 'professional certification' under several consent decrees with the US Environmental Protection Agency (EPA) and Department of Justice (DoJ). In a similar way, EPA lists a number of stormwater, ecological, and erosion and sediment control courses on its website that it delivers itself and also are delivered by a range of other certified trainer organizations.[25] We've seen how the NGICP certifies both its trainers and also, after they have passed a test, its trainees,[26] while the Water Environment Federation's Knowledge Center also lists stormwater and erosion and sediment control training providers.[27]

And, of course, the world of professional training has its own certification programmes – especially for training evaluation. If you get really into your training, you can become a Kirkpatrick Certified Professional for the first four levels of evaluation and a Certified ROI Professional® (CRP) for Level 5 evaluations.[28]

Would it work if your organization were to recognize training offered by other parties, such as by certifying the appropriate vocational training institutions, private training providers, professional associations and so on?

Certification of products in your guidelines   Ongoing innovation is a wonderful feature of environmental management: people are always dreaming up new and better ways of doing things. But with 'fail faster', the mantra for successful innovation, we run the risk of adopting things that later on turn out to be not such a good idea.

This raises a problem with technical guidelines: when do you add a proprietary product to your guideline? Most public agencies are reluctant to specifically endorse a commercial product and will instead specify a performance outcome or a generic class of products. So what happens when a commercial provider comes up with a bright idea? How much evidence is enough to justify adding it to a guideline or collating a 'list of alternative measures' whose use must be justified on a case-by-case basis?

After several years of engaging with the stormwater industry, Auckland Council developed a protocol for assessing proprietary stormwater treatment devices.[29] It gave clear and detailed instructions for vendors on how to assess and report on the performance of their device and for council staff or other evaluators on how to evaluate these reports. It set up two pathways for vendors to prepare the performance claim against which the device is evaluated: one based on existing international and/or local trials; and the other, on setting up a local trial to provide such evidence. The protocol set out in detail the methods of ensuring that evidence was provided in a consistent way. Certificates were to be valid for five years and then renewed, with certificates published on the council website, along with an overview of the evaluated performance and capabilities of the device and any specific provisions. The aim was to provide a process that is transparent, consistent and timely – and it was interesting that both council staff and the commercial sector expressed very similar hopes and concerns about the protocol. The protocol has not yet gone live at the time of writing, so watch this space.

Are you working in an environmental field with rapid innovation? How would this approach work for you and all of the public and private stakeholders involved?

Certification of infrastructure or buildings   Green buildings, green roads and green infrastructure can be certified, and are increasingly required by clients in the public and private sectors.[30] Things are changing so rapidly that I recommend you run a comprehensive search online for 'green [whatever is of interest to you]' as you develop and update your environmental training programme.

Accreditation   Like 'certification', the term 'accreditation' is also used to denote some form of more-or-less formal approval, recommendation or authorization of a person, agency, product or process.[31] The terms seem to be used more or less interchangeably, with perhaps more focus on accreditation of people and agencies than of products or processes. Regardless of the terminology, it's clear that organizations can set up their own certification or accreditation systems and then certify or accredit other agencies, as well as people, products or processes.

Which existing certification or accreditation system would work for you? Would it work for you to create one of your own? Remember that the 'do nothing' option can also be valid.

Qualifications

People with university degrees are often happy simply to get CPD points for attending training. If they want to get a higher degree or a specialist diploma, they are usually motivated enough to organize this for themselves. However, as environmental work becomes more specialized and very targeted training is developed to foster these skills, there will be more demand for narrow and deep skill sets to be recognized by properly authenticated micro- or other credentials.

Qualifications from technical colleges or industry training bodies are usually more suitable for people who are unqualified and want to gain a qualification, those who are undergoing further education or those who want the satisfaction of achieving a particular specialist skill. Gaining a qualification can give a huge sense of achievement for people with no formal qualifications who are nonetheless very skilled operators who take pride in their work – often the unsung heroes of front-line environmental work.

However, existing qualifications may not cover environmental performance to the standard your organization needs or wishes. In that case, you can work with these institutions to develop appropriate content and resources, as in the example of the Digger School in Chapter 2 and the following case study. This can be a very good option, especially when supported by e-learning in trades and professions in which such large numbers of people need environmental training that government bodies and companies don't have the resources to deliver at the necessary scale.

> **Case study: sewage overflows – a learning team develops a vocational training qualification**
>
> Many years ago, I was tangentially involved with the development of a vocational training qualification for managing dry-weather sewage overflows. The trainees were the staff and contractors of sanitary utilities, as well as the engineering and environmental consultants working with such infrastructure. The council was

concerned about the community and environmental outcomes of the overflows and got together with the utilities to work out what to do. Among the solutions, which included more storage, alarms and inspection of the infrastructure, as well as a focus on preventing the causes of the blockages, they agreed that prompt on-the-ground response was needed and that the contractors – the first responders – needed very good training.

In line with the partnership principle, a working group was formed of representatives from the relevant industry training organization, council, utilities, contractors and specialist consultants. The first piece of work was to fully understand the situation and define what best practice response looked like. All parties thoroughly reviewed this report, and the working group then defined the training needs and learning outcomes. The industry trainer was able to define the level of qualification that people completing the training would receive and translated the technical report into the formal structure needed to describe and deliver the training outcomes.

This collegial and collaborative process showcased the complementary areas of expertise of everyone round the table and resulted in the efficient production and delivery of a much-needed training resource.

These days we'd call the working group a 'learning team'. A learning team aims to ensure that the people who do the hands-on work are involved in the development of the training – building on the avatar process described in Chapter 5. Greg Dearsly says that the groups needing to be involved include, for example:

- managers responsible for organizational success, contractual obligations and logistics;
- front-line workers responsible for on-the-ground activities; and
- stakeholders from, for example, an industry association, the council or a community group.[32]

Greg notes that the best results emerge when these different groups come together as a learning team and where:

- the outcomes are clear, understood and agreed;
- the team process is unstructured and organic;
- open discussion in 'learning (or discovery) mode' can reveal the non-linear complexities of how everyday work is done; and
- once the 'work as done' is fully understood, the team can move into 'problem-solving mode' to deliver the desired outcomes.

As Greg asks: 'Many businesses say they are a learning organisation and that's a great aspirational goal, but what is actively being done to achieve this?'[33] His article makes a real contribution to 'learning as a deliberate action', which fosters the development of learning organizations.

Licensing or registration

Licensing or registration is compulsory in some professions such as law, medicine, architecture and engineering, in which you must be fully qualified to practise. In most such cases, practising certificates can be withdrawn on the grounds of professional misconduct. This is often a system backed up by legislation, which can include legal sanctions against those concerned. Such systems are much more formal and enforceable than an environmental certification: most of us can continue to carry out our work without undergoing recommended environmental training, though in some places we may find it increasingly difficult to win work without it.

Summing up

Training often starts in an ad hoc, informal way and gradually becomes more formalized, so any time is a good time to look at your programme with new eyes – remembering that 'start simple' is always good advice. Some straightforward steps towards a comprehensive training framework might be summarized as follows.

1. Develop a simple, but robust, assessment of learning outcomes from training delivery.

2. Work with the relevant professional associations to deliver workshops that enable their members to gain CPD points.

3. Work with the relevant vocational and adult training institutions and industry representatives to develop qualifications based on workplace competencies and deliver the training to professional tradespeople and the regulatory staff who inspect them.

 ACTION PLANNER

What possibilities arise for you from these ideas? Check that you've captured them all in Action Sheet 6.2 in the free Action Planner that accompanies this book.

6.4 ASSESSING THE CONTRIBUTION OF TRAINING TO YOUR PROGRAMME OUTCOMES

> One of the great mistakes is to judge policies and programs by their intentions rather than their results.
>
> Milton Friedman[34]

The whole idea of the environmental programme of which our training is only one part is to produce the desired third order environmental and associated outcomes illustrated in Figure 3.5.

Training contributes to desired outcomes by giving trainees the information and skills they need to perform to the desired environmental standard, by encouraging and requiring their supervisors and managers to support the application of this learning in the workplace, and by encouraging managers to map the training to business outcomes. Over time, the business for which your trainees work will hopefully be developing a good understanding of how better environmental performance makes a positive difference to its bottom line.

It is then up to the environmental management body responsible for measuring the state of the environment to identify any positive changes over time and work out what changes can be attributed to training and what can be attributed to other management interventions, as well as what their relative contributions may be. This may be difficult or even impossible to 'prove' beyond reasonable doubt, but an attempt must be made.

Essentially, training is a method set out in your environmental management programme (which is a first order outcome) that will achieve the desired changes in practice (second order outcomes) that will produce the desired environmental states you want to achieve (third order outcomes). As is the case for the other environmental management tools in your programme, such as research, policy, regulation, inspection and enforcement, the cost-effectiveness of your training needs to be monitored and evaluated.

Figure 6.4 shows how these outcome areas – learning, workplace performance and business performance – all need to be aligned with each other, across the actions and responsibilities of different parties, to produce the desired environmental outcomes and how breaks or misalignment between them will slow progress towards the environmental outcome. When aligned, however, they will together produce the Levels 4–6 or perhaps even Level 7 training outcomes that will deliver the third order environmental outcomes you want.

Successful environmental training depends on a suite of individual, workplace, business and environmental performance drivers. If training is to make a difference, then the wider industry must have a very good understanding of the business and wider benefits of good environmental performance. The partnerships set up at the start of your programme will foster this understanding.

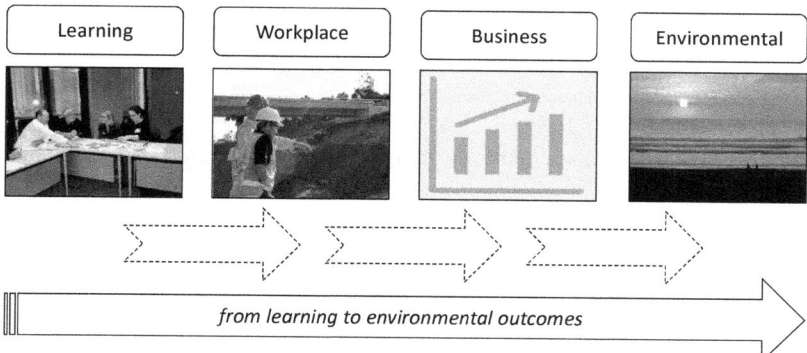

Figure 6.4 Aligning learning, workplace, business and environmental outcomes

6.5   TIPS FROM SOME EARLY EVALUATIONS

> The only real mistake is the one from which we learn nothing.
> Henry Ford, founder of the Ford Motor Company[35]

The ultimate test is whether the training makes a difference to the state of the environment.

Given how little we knew when we set up the Auckland erosion and sediment control training programme, it's amazing how well we evaluated its effectiveness. This is very much to the credit of the programme managers listed in the acknowledgements. I believe that the industry has started to progress towards some elements of Alastair Rylatt's Level 6 and Level 7 measures of sustainability on some topics – albeit with the occasional hiatus. The following are some of the things we did for Levels 1–5 evaluations.

- We handed out Level 1 'smile sheets' at the end of each workshop and, although they weren't worded as carefully as I'd word them now, our trainees almost universally rated the workshops at 4 and 5 on a five-point scale (1 = bad; 5 = good). We always took on board trainees' good suggestions for improving the workshops, fully debriefed with our client, and updated the content in line with new scientific, technical and policy developments.

- We started by setting post-workshop Level 2 tests of learning and would issue attendance certificates when they were returned, but lacked the administrative capacity to keep it up. However, annual surveys of the industry indicated that up to 90 per cent of respondents agreed that the training component of the programme had dramatically raised their awareness of the potential adverse environmental effects of uncontrolled erosion and sediment runoff.

- Level 3 performance assessments were of three main types: desk-based, on-site and research-based.
  - Desk-based assessments by council staff indicated that the quality of applications for environmental permits had steadily improved.
  - Weekly site inspections showed a steady improvement in the standard of construction, operation and maintenance of site controls.
  - Ongoing research into improved erosion and sediment controls showed a steady increase in the proportion of mobilized sediment able to be cost-effectively retained on construction sites.

- Level 4 evaluation of the business impacts on the individual firms whose trainees attended our workshops was beyond our capability at the time, but an indirect measure that came across even then was the increased focus by agencies commissioning works on the environmental scoring of erosion and sediment controls in bidders' non-price tender attributes (see Chapter 3). This heightened awareness of environmental performance also led to environmental controls being priced on a scheduled basis instead of lump sum, with the important result that contractors were paid for their environmental work instead of squeezing in bare minimum controls more or less at their own cost.

- Level 5 financial ROI results for the programme as a whole were also well beyond our grasp at that time, but anecdotal evidence from the industry suggested that if ROI were measured for individual companies (for example the cost of the workshop fee and time off-site for trainees compared with the financial benefits of improved environmental controls on sites), results could include monetized measures of improved productivity as a result of better risk management, less storm damage to sites and better success rates in tendering for work as a result of improved non-price attributes. This would more than justify the companies' financial investment in sending staff to the workshops. From the council's point of view, it would also enjoy the reduced costs of processing improved environmental permits and less legal enforcement.

⊘ KEY OUTPUTS

At the end of this stage, you will produce a report to your managers, funders, partners and key stakeholders on your training evaluation results and what actions you will take in light of its findings.

 ACTION PLANNER

Revisit Action Sheet 6.1 in the free Action Planner that accompanies this book to see if you can add any more ideas about how to evaluate the effectiveness of your training.

NOTES

1 This quote is usually attributed to Winston Churchill and wrongly so, according to Churchill scholar Richard Langworth – see https://richardlangworth.com/quotatioins – but the original source is unknown: see www.brainyquote.com/quotes/unknown_135256 [both accessed 15 August 2019].

2 Katie Paine, quoted in Karen Scherer (1999) Market the event – then evaluate your success says industry guru. *New Zealand Herald*, 11 February. Find out more about Katie Delahaye Paine's excellent work at http://painepublishing.com/blog-2/ [accessed 25 October 2019]. Katie specializes in measuring the effectiveness of strategies used in the world of PR research and evaluation, and her thinking is very helpful and accessible.

3 Cited, but not referenced, in Dr Madan Manandhar (2007) Training for change: think performance need first not training. *Administration and Management Review* 19(2), p. 1. Available at www.nepjol.info/index.php/AMR/article/view/898 [accessed 25 October 2019].

4 (1) David Kirkpatrick (1998) *Evaluating training programs: the four levels.* Berrett-Koehler. Find out more at www.kirkpatrickpartners.com/Our-Philosophy/The-Kirkpatrick-Model [accessed 5 June 2019]. (2) Dr Jack J. Phillips (1997) *Return on investment in training and performance improvement programs.* Butterworth Heinemann. Find out more at https://roiinstitute.net/ [accessed 5 June 2019].

5 Clare Feeney (2018) *ROI on environmental training: how to measure the full financial return on investment of your environment and sustainability training.* Workbook for a one-day workshop for environmental professionals.

6 E. Stefanakis (2002) *Multiple intelligences and portfolios.* Heinemann, p. 9, referenced in the extremely useful discussion of assessment at https://sites.google.com/site/assess4learning/assessment-defined [accessed 5 June 2019].

7 Find out more about Johann's and Eddie's sustainable building work at www.bernhardtarchitecture.co.nz and www.equinoxdesign.co.nz [both accessed 5 June 2019].

8 Dr S. Geertshius, C. Feeney and G. Ridley (2012) *Stormwater industry training: investigating certification and accreditation opportunities.* A report prepared by Environmental Solutions Ltd, Environmental Communications Ltd and RidleyDunphy Environmental Ltd for the Auckland Council, May. I am very grateful to the Council for permission to use and adapt material from that report.

9 Also attributed to Rob Siltanen and Lee Clow for Apple's 1997 'Think Different' campaign and to John McAfee by www.brainyquote.com and www.goodreads. com [both accessed 15 August 2019].

10 See Michael Porter's work at www.hbs.edu/faculty/Pages/profile.aspx?facId=6532 and Peter Senge's at https://mitsloan.mit.edu/faculty/directory/peter-m-senge [both accessed 20 May 2019].

11 (1) Ioannis Ioannou and George Serafeim (2019) Yes, sustainability can be a strategy. *Harvard Business Review*, 11 February. Available at https://hbr. org/2019/02/yes-sustainability-can-be-a-strategy [accessed 30 May 2019]. (2) Jeff Kauflin (2017) *The world's most sustainable companies 2017. Forbes*, 17 January. Available at www.forbes.com/sites/jeffkauflin/2017/01/17/the-worlds-most-sustainable-companies-2017/#2290ee3c4e9d [accessed 30 May 2019].

12 S. Collins (2008) Taggers' target begins fightback. *New Zealand Herald*, 30 October. Available at www.nzherald.co.nz/wellington-city-council/news/article.cfm?o_id=240&objectid=10540292 [accessed 5 June 2019].

13 S. Collins (2008) Employing prisoners pays. *New Zealand Herald*, 7 November. Available at www.nzherald.co.nz/nz/news/article.cfm?c_id=1&objectid=10540293 [accessed 5 June 2019].

14 K.J. Moore (2012) Investing in ex-cons: small loans may yield big opportunities for social reintegration. *The Futurist*, Sept–Oct, pp. 12–14.

15 Dr Carol S. Dweck (2017) *Mindset: changing the way you think to fulfil your potential*. Robinson, p. 141. Dr Dweck is Lewis and Virginia Eaton Professor of Psychology at Stanford University.

16 Dr S. Geertshius, C. Feeney and G. Ridley (2012) *Stormwater industry training: investigating certification and accreditation opportunities*. A report prepared by Environmental Solutions Ltd, Environmental Communications Ltd and RidleyDunphy Environmental Ltd for the Auckland Council, May. I am very grateful to the Council for permission to use and adapt material from that report, and to Susan for helping me develop my thinking in this area.

17 Emmanuel Olaoye and Stuart Gittleman (2014) Effective training a weak link in many compliance programs: survey. A guest blog by Compliance Complete. Available at http://blogs.reuters.com/financial-regulatory-forum/2014/08/13/effective-training-a-weak-link-in-many-compliance-programs-survey/ [accessed 30 May 2019].

18 Mark Iafrate (2017) Digital badges: what are they and how are they used? Available at https://elearningindustry.com/guide-to-digital-badges-how-used [accessed 6 June 2019].

19 See www.businessdictionary.com/definition/certification.html [accessed 5 June 2019].

20 See e.g. www.ieca.org/IECA/Education/IECA/Education/Education. aspx?hkey=45eb2999-5d38-43a5-b129-c3e7d9ad3ab5, envirocertintl.org/cpesc/ and www.eianz.org/institute-programs/certified-environmental-practitioner-scheme [both accessed 12 August 2019].

21 See https://wiki.answers.com/Q/What_is_the_definition_of_certification [accessed 5 June 2019].

22 Roger Bannister and Clare Feeney (2009) Success by collaboration: the Auckland Regional Council's Erosion and sediment control training programme. A paper presented at the 33rd International Association of Hydraulic Engineering & Research (IAHR) Biennial Congress, 9–14 August, Vancouver, BC.

23 Information about this programme can be found at www.bae.ncsu.edu/workshops-conferences/stormwater-bmp/ [accessed 5 June 2019].

24 See www.stormwaterone.com/ [accessed 5 June 2019].

25 See www.epa.gov/compliance/national-enforcement-training-institute-neti-elearning-center www.1.usa.gov/11akWFw [accessed 5 June 2019].

26 See https://ngicp.org/program/ [accessed 30 May 2019].

27 See http://knowledgecenter.wef.org/StormwaterONE and http://ieca.learnercommunity.com/wef [both accessed 30 May 2019].

28 See www.kirkpatrickpartners.com/Training-Events/Four-Levels-Certification-Bronze and https://roiinstitute.net/roi-certification/ [both accessed 30 May 2019].

29 Grace Wong, Judy Ann Ansen and Elizabeth Fassman (2011) *Proprietary device evaluation protocol*. Auckland Council Guideline Document GD003. I am grateful to the Council for permission to use and adapt material from that report.

30 Many countries have their own local green building certifications systems. Also see the not-for-profit Greenroads Foundation at www.greenroads.org/publications and the Infrastructure Sustainability Council of Australia (ISCA) www.isca.org.au/ [both accessed 12 August 2019].

31 See www.thefreedictionary.com/support [accessed 5 June 2019].

32 Greg Dearsly (2019) Using learning teams to enhance organisational learning and worker engagement. *Revolve*, the magazine of the Waste Management Institute of New Zealand, 173, pp. 20–23. Greg is the owner of First 4 Safety Ltd and has been providing business with workplace safety and health advice for 18 years.

33 Ibid., p. 23.

34 Milton Friedman, in an interview with Richard Heffner, on *The Open Mind* on 7 December 1975: see https://en.wikiquote.org/wiki/Milton_Friedman [accessed 15 August 2019].

35 Widely attributed, but not traceable, to Henry Ford. See www.goodreads.com/quotes/106813-the-only-real-mistake-is-the-one-from-which-we [accessed 15 August 2019].

7

Managing your training programme for success

> The road to success is always under construction.
>
> Chinese proverb[1]

Administration is the ugly duckling in the Success Framework. In Figure 3.1, it's called re-sourcing and endorsement, but it boils down to good management. It's not as immediately appealing as the other six elements of success, but, like each of those other elements, its absence will weaken the effectiveness of your training programme.

7.1 TIME

> Hofstadter's Law says it always takes longer than you expect, even when you take into account Hofstadter's Law.
>
> US Professor of Cognitive Science Douglas Hofstadter[2]

Don't underestimate the time it takes to keep your training programme going. If you think you can do it all on your own, take a look at the work involved in organizing just one training workshop – let alone the other moving parts in your training programme – out-lined in section 7.4.

Regardless of the scale of your training programme, you will find it invaluable to prepare a management plan with even just a single page each on:

- a funding plan and budget;
- a partnership plan;
- an engagement and communication plan;
- workshop management procedures; and
- an evaluation plan.

Preparing these plans, along with the work you have done in the preceding chapters and the free Action Planner that accompanies this book, will give you some idea of the time you will need to allocate for your training programme. If you can, budget for some administrative assistance for the tasks I'll be describing.

 **ACTION PLANNER**

As you read this chapter, use Action Sheets 7.1–7.3 in the free Action Planner that accompanies this book to track your thinking on the things you'll need to put in place to support the efficient management of your success environmental training programme.

7.2 MONEY

We don't have much money to do this so we're going to have to THINK.

Ernest Rutherford[3]

If you really want to track the effectiveness of your training programme to keep a great business case in front of your managers, then you need to work out what to include in your budget.

You'll need to track both income and expenditure in your annual budget. You may be fortunate enough to be able to fund all or most of your training programme or to run it on a fully commercial model, but most of us will need to consider a mix of funding options, including:

- attendance fees;
- sponsorship from government and/or business bodies (see Chapter 5);
- in-kind support for costs of venues, printing and the like;
- certification fees; and
- discounts from training providers if economies of scale or guaranteed work make this commercially viable.

Your managers will most likely ask you to prepare an annual budget, and tracking income and expenditure will be essential for fully evaluating the return on investment (ROI) of your training.

If you decide to evaluate the ROI of your training, you will need a way of tracking the value of returns accrued across the six (or seven) capitals. Things such as higher skills (human capital), community development (social and relationship capital), indigenous development, ecosystem services and so on are not traditionally accounted for in book-keeping systems. You will also need to count things such as staff time, sponsorship and in-kind contributions. In-kind contributions could include things such as offering a venue, allowing trainees to visit a workplace, or suppliers demonstrating new methods or equipment at a field day.

Full cost accounting of social, indigenous and environmental resources is a contested area, but we don't have to believe that the numbers are 'real' in an absolute sense and they do make people think. After all, how will we know how much to spend on an environmental initiative if we can't place that expenditure within the wider context of the costs of environmental, social and other harm and the value that can be gained by improving outcomes in those areas?

7.3 TECHNOLOGY

Technology is best when it brings people together.

Matt Mullenweg[4]

You need technology to manage your training programme and to provide it with a web presence.

Stakeholder database, CRM system or LMS

Every training programme needs a comprehensive stakeholder database system, because you'll end up having lots of information collected for a lot of related purposes. You can track your training on a spreadsheet or customer relationship management (CRM) database, but if your human resources (HR) personnel have a learning management system (LMS), then ask them to include your training programme in it. An LMS is a software application for the administration, documentation, tracking, reporting and delivery of educational courses, training programmes (including e-learning), and learning and development (L&D) programmes. There are three main LMS user groups (some of whom may be the same people):

- **administrators** – the people setting up and configuring the LMS platform from which online training content and other information is distributed and updated;

- **instructors** – the people preparing the training and accessing the learners' progress; and

- **trainees** – the people doing the learning.

Learning management systems are becoming increasingly sophisticated, with inbuilt authoring tools allowing trainers to develop online training materials without additional third-party software, and many more bells and whistles. The benefit of even the most basic LMS is that it will allow you to track the training of external and internal trainees, manage their learning assessment results, and enable reporting and feedback.

The information to put into your database should help to you manage, for example:

- your trainees – start by collecting contact details of who registered for the class, who attended (if there was a last-minute substitute), who took any test, who passed, who had to re-sit and who received the certification of attendance or other recognition of training, plus all related dates;

- your internal and external partners, plus links to meeting dates, agendas and minutes;
- scores or other results from site inspections;
- people and organizations who have won awards or been prosecuted;
- workshop evaluation summaries;
- learning assessments;
- evaluation summaries of other events;
- workshop and field day sponsors;
- good training venues and costs; and
- good service providers, such as caterers and bus companies.

This information will provide basic data for the purposes of evaluation and follow-up. Make your databases as comprehensive as you can, while remembering that people are mobile and you will be updating it regularly. The value of this information will become progressively more evident as you use it in combination with other information in your logic model, such as compliance monitoring data and environmental outcomes.

Online presence

Your training programme needs an online presence so that people who are looking for something or who have heard a vague mention of environmental training can find it. It's not always easy to get a dedicated page set up on your organization's website and to build all the links to it from related pages – but it is very important to do so.

Closed groups on social media platforms such as Facebook are increasingly common for initiatives like training, so a branded and moderated group could work well if you can get approval for it.

On your page or site, consider the need for:

- links to other relevant pages – especially to where your technical guideline is located;
- a link to find out more about the training (content, target audience, dates, fees, locations and so on), including how to register or make further enquiries;
- testimonials from past trainees;
- a click-through button to enable people to sign up to the mailing list; and
- minutes and/or presentations from industry meetings or events.

What other information do you think the industry, other environmental agencies and communities might find useful?

7.4 SYSTEMS AND PROCEDURES

> A good system shortens the road to the goal.
>
> Orison Swett Marden[5]

Administration: it's the bane of our lives, isn't it? Yet a few key systems will make things so much easier for you as you run your training programme.

Document what you do

Helping a new person to walk into the administrative role when the previous incumbent isn't around to explain simple administrative tasks is hugely time-consuming. I'm not suggesting that you develop a massive procedures manual, but even basic notes about key tasks and a flow chart, with a checklist for workshops and other events, will be helpful. Good internal knowledge management systems, along with strong external partnerships, will also help to retain organizational learning in both the public sector, which is vulnerable to electoral and legislative change, and the business sector, in which firms can also undergo major organizational changes.

Tools such as Jing or Snagit let you record active screenshots and voiceovers to build an electronic systems manual – a great shortcut. Use these and other tools to help you to quickly and easily document how to carry out key tasks, from booking buses and venues to updating the all-important database.

Some of the things you can compile in a checklist, set of instructions or a mini-procedure manual include how to:

- find and book venues;

- sort out equipment such as a data projector, whiteboards and so on;

- organize catering (a very important task – people always comment on the food!);

- arrange transport for any field trip;

- obtain health and safety clearances if needed for the field trip;

- prepare booklets and handouts;

- market the training;

- complete registration forms and procedures;

- draft a letter of confirmation noting the time, date and place of the training, as well as details such as parking and what to bring on the day (for example personal safety equipment, a scientific calculator and so on);

- attract advertising to fill vendor or sponsor spaces;

- prepare participant name plates or badges (for the benefit of the trainers);

- design and deliver attendance certificates; and

- get people to sign up to your mailing list.

Take care with dates. Be vigilant for public holidays, school holidays, daylight saving changes and the like. I avoid Mondays and Fridays, too, so that people can start and end the week at work.

 **TOOLBOX**

Go to www.ESST.institute/Success/Toolbox to download some forms and templates to help you to manage your training programme and workshops.

Case study: Tania's toolbox

I was delighted, during one workshop, to be handed a toolbox – literally – by one of the many 'organizers par excellence' with whom I've been privileged to work over many years of training. Tania's toolbox was a classic, bright red, metal toolbox that opened up to reveal compartments and trays. It contained spare pens, pencils, sticky stuff to fix flip charts to the walls, sticky notes, scissors, tape, a small first aid kit and all sorts of other useful things – even including four sets of different coloured flip chart and whiteboard pens, all full of ink.

Now that's organized.

What would you put in your workshop toolbox?

Set up an image and media filing system

An image and media repository is another must-have. One of the things everyone struggles with when starting a new training programme is gathering a good collection of local shots: pretty and relevant scenic shots, and what we call 'the good, the bad and the ugly' images of workplace performance and environmental effects.

Start collecting photos as soon as you can and if you don't have any pictures of good practice in your locality, ask to use some from elsewhere until you do. Portraying best (and worst) practice with images is an invaluable performance improvement tool, so actively encourage staff and stakeholders to add to your collection by taking photos during site management and compliance monitoring.

Store these in one place (you will need to keep reminding your staff and consultants to upload their photos to a common archive) and, for ease of retrieval, record the date and site (and/or file number) in the title. If you use the date format YY/MM/DD, your images will automatically sort themselves in order of date, for example '2019_04_25_Rocky_Creek_badpic' or '2019_09_27_Rocky_Creek_goodpic'. Use subfolders to sort images into relevant categories for topics such as river basins, industrial spills, sediment or other topics relevant to your training.

Adding your images to the archive in PowerPoint format means you can make sure that each image is accompanied not only by date, time and place details, as well as the name and role of the person who took the photo, but also by a more detailed explanation of what is shown and links to other relevant files. Whenever I think I couldn't possibly forget that image and fail to include the information, I invariably come back to it some years later and find myself at a total loss. Including key information in the 'Notes' format is invaluable for learning and will help new staff to come up to speed. It will also help your trainers, if they are not the people who took the photos.

Keep a detailed electronic file of any media coverage. People love to read about their sector in the news, so start making a collection of relevant stories – video, audio or print media – to use in your training. Also collect the stories about environmental infringements or prosecutions: somehow, it seems more 'real' to trainees to see a newspaper clipping or hear a radio announcement about a recent event than simply to be told about it by the trainer. A picture is worth 1,000 words, and media coverage can inform, sting and praise your target sector, all of which encourage good environmental practice.

If you want to reproduce media images or clippings in your presentations and workbooks, you are likely to need to gain copyright approval to do so. Contact your communications staff or the media concerned to find out more about this.

7.5 COMMUNICATION AND MARKETING

The great enemy of communication, we find, is the illusion of it.

William H. Whyte[6]

You'll need a range of communication channels and contents for the full spectrum of people, organizations and groups involved in your environmental training programme. You'll need a communications plan to keep in touch with them in a systematic way, even if it's just a one-page annual calendar.

Internal partners

Given that your internal partners are likely to be scattered all round your organization, you need a plan for keeping in touch with them. Talking is great! If you don't sit near each other, why not stroll by their desk or organize a regular catch-up? Add them to your email database so that they get regular programme updates. Put them on the invitation list for regular programme meetings. For more senior people and elected representatives, write quarterly reports on progress in all aspects of your wider programme, as well as the training component – and say thank you! This will all help to keep your programme fresh in their minds.

External partners

For external partners, a simple 'thank you' goes a long way. They are usually busy people who give considerable time and thought to your programme and fund the time of their staff to do the same. How will you document and acknowledge their generosity? How and how often will you communicate with them?

- Government bodies might run a regular forum with external partners. It's an invaluable way of discussing issues of concern and constructive opportunities.

- Companies doing their own internal training should involve their HR and communications departments, as well as senior management and shop-floor representatives from the department undertaking the environmental training. Attracting their interest and support will also be very helpful for calculating the ROI on your training.

- All past trainees can opt in to receive regular newsletters.

Ongoing communication with your industry is vital, but it takes time to put together even three or four newsletters a year. Make sure you allow for enough time and resources to make good use of the communication channels you choose.
Other options include:

- webinars;

- seminar evenings;

- annual field days, with presentation, displays, demonstrations and more; and

- uploading presentations, images and other information to your website.

Marketing

Marketing your training needs a compelling sales pitch, multiple channels and repeated releases. Who should attend your training and why? What are their pain points? What will they gain? Refer to any continuing professional development (CPD) points or other recognition they can claim for attending. As well as your newsletter, offer articles from the newsletters of key partner organizations. You know your programme is getting traction when people call or email you asking how to register: make sure the contact details through which they can do this are on all of your communications.

Conventional and social media have a role, and environmental matters are increasingly newsworthy. If you have a communications team in your organization, they will have more great ideas for you to follow up.

 ACTION PLANNER

Use Action Sheet 7.2 in the free Action Planner that accompanies this book to jot down some notes about the frequency and timing of key communications and marketing dates. Include your workshop dates and registration dates, if you know them, as well as public holidays and other key dates to plan around, so that you can put together a calendar.

Environmental awards

Once your programme has got some traction and credibility, consider setting up an environmental award or introduce one into an existing awards programme. Such awards add a very useful element of competition into the game. Having a mix of judges from indigenous peoples, sector representatives, regulators and more will help to foster understanding and appreciation across the sector.

If you're thinking of introducing an award, revisit the considerations in Chapters 1–3: what is the issue of interest, and how can growing green jobs and industry capacity and capability help to address it? These will help you to define specific criteria for your awards. Generic criteria could include, for example:

- benefit to the environment;
- community participation;
- significance to the local community;
- leadership and innovation;
- sustainable management practices;
- savings of time and money;

- project resilience; and

- overall sustainability of the project.

Simply reading the entry form can make people think very hard about their environmental performance, even if they don't enter! Awards are usually a highlight of the black-tie functions at annual conferences and also liven up less formal events such as annual field days. All of this is a great way of raising the sector-specific profile of good environmental practice.

7.6 PROGRAMME REVIEW

> Building enterprises capable of continually adapting to changing realities clearly demands new ways of thinking and operating. So do the sustainability challenges [we face], in many ways the archetypal organizational learning challenges of this era.
>
> Dr Peter Senge, MIT Sloan[7]

An annual review and, every three to five years, in-depth reviews will enable you to pick up positive and negative trends across all seven elements of your training programme. You need to stay in touch with and support the people responsible for these other parts of the Success Framework because they are generating new information all the time, and you need to know about it to keep your training current and relevant.

Ongoing research, monitoring and evaluation

Evaluation of your environmental training may identify the need for more research. This may be comparatively simple studies of the effectiveness of environmental controls, with which the industry and perhaps a university student may be able to help you – and I've seen some extremely useful results come out of such studies. It may be more sophisticated studies by specialist research bodies or it may involve a review of existing monitoring programmes and the identification of wider environmental and industry issues. Such work needs funding and may have implications for other people's budgets. Your in-house monitoring or auditing teams may be able to tie your work into their existing research or help you to set up small studies yourself.

Build links with indigenous community monitoring or citizen science groups: they may have various projects that could be relevant to you. And consider further building research-related capacity and communication in your jurisdiction by involving learning and research agencies.

Keep up your own research as an environmental professional and a trainer. I try to stay up to date across business, the environment, economics, ethics and risk, as well as training. I subscribe to several magazines, my top favourites being *New Scientist* and the *Harvard Business Review*, and I also get magazines from the many professional associations to which I belong. I subscribe to e-newsletters across all those topics, albeit with period purges as they multiply. Make the most of networking opportunities, including conferences, monthly meetings and the like – especially those through which you meet professionals from other fields. We're all interconnected!

Policy, regulation, compliance and enforcement

A key purpose of environmental monitoring and evaluation is to assess how well policies and their methods address issues and maintain or enhance values. You therefore need to work closely with your policy, regulation, compliance and enforcement teams as well as your internal and external partners to assess how well your training is working. In some cases, this may mean expensive and time-consuming changes to policies, plans, rules and regulations, but in many cases this won't be necessary. More targeted and streamlined implementation of the existing framework is often sufficient.

Your technical guideline

The first edition of a technical guideline will inevitably have wrinkles that need ironing out and it will need to be updated as new and better practices emerge. Changes need to be balanced with the fact that a guideline, especially if associated with a statutory process, can't be a constantly moving target: the industry needs certainty about the performance standards it is obliged to meet. But when site inspections and the industry's own experience – and what will become its increasing expertise – indicate that a change *is* needed, an update will usually be welcomed. Work with your industry partners, keep amendments to the minimum required and publicize any changes, along with any new measures, via as many different communication channels as you can.

Your training and your trainers

Chapter 3 suggested a moderation process to ensure that different trainers, even on different topics, maintain a consistently high standard of training delivery, currency and relevance. In this regard, the results of the evaluations described in Chapter 6 will be very helpful. Whether you have a single trainer or 100 trainers, hold a half- or one-day annual excellence workshop with them to go over the training evaluation results, the overview of the wider environmental programme, recent developments and the like. Have a session in which you all share the professional development that you and your trainers have undertaken to grow both technical and training expertise.

Why not invite a local training professional to give a stimulating presentation with a question-and-answer and discussion session to inspire you all? You can then wrap up with a working session on what aspects of the training might be dropped, added and improved – and, of course, a tasty morning or afternoon tea or lunch will be a great thank you to everyone.

Your partners

Whatever thanks and recognition you give others, give double to your partners – especially your external partners. We don't all work at the scale of the Chesapeake Bay Program, but offer your partners informal networking and professional development opportunities, as well as a similar annual briefing to that which you deliver to your trainers. Invite your partners to speak at annual field days or conferences, and if they need to step down from their partnership role, give them a certificate of recognition. It's these people that keep you credible within your industry – so look after them well.

Continued resourcing and support

It's a boring, but necessary, truth that you *must* keep track of your resource needs and usage, and document what you spend time and money on. Make it clear how each investment supports your work and contributes to your programme outcomes. This will help you to justify your costs and, if necessary, work out where you can make savings that won't overly compromise the programme's effectiveness in the event that budget cuts are needed. This is something we all face from time to time. Such documentation will also help if you or other parties want to set up a similar programme.

More importantly, if you want funding for your programme to continue, work with all the foregoing elements to maintain and grow the enthusiasm of the people whose endorsement you need. And if you're worried about a change at the top, consider the following case study.

Case study: from critic to convert

More years ago than I care to admit, when I was a very junior water resources scientist for the regional environmental regulatory body, our local body elections delivered us a chair who'd been elected on a platform of curbing the regulator's excessive powers. He'd been on the receiving end of some legal compliance discussions when we'd found him using his bulldozer on a popular bathing beach – a beach that he saw as a cheap source of sand for his building company. He decided that our intervention was a step too far for democratic freedom.

We were all very nervous about his election, but our senior managers proceeded with the usual induction programme for our newly and re-elected political representatives. This included an afternoon workshop on the legislative context for the organization and the work we did to meet those legal requirements, as well as two half-day tours of the region to look at the environmental values and land-use activities we were there to manage.

Over the course of the first year of his term, this councillor asked a lot of hard questions that kept our managers very much on their toes. But, by the end of his three-year term, this sometimes intense engagement meant that he had become one of our most vocal and articulate supporters, and he was re-elected for a second term on that basis.

Never give up hope!

Holding the big picture and the long view

Over time, I've seen some initiatives as diverse as solid waste minimization, industrial resource efficiency, pollution prevention, school education programmes, plant and animal pest control, and urban streambank protection – to name but a few – fall by the wayside as a result of changes in organizational structure, senior staff or elected representation. Even without these hazards, it's all too easy for agencies to withdraw or reduce support because a programme is thriving only for it to starve a short time later. Successful programmes also run the risk of 'hostile takeovers' from other parts of the organization wanting some of the success.

It's consequently very important to maintain support for your programme not only within your organization, but also with your industry partners and the wider community. Others can speak up to keep it going when you can't. Staff of environmental agencies are often asked to give talks to various community and professional groups, and that's a brilliant way of freshening up your thinking and keeping the public informed about the work you do and what services they get for their rates. Entertainment and education go together well: I don't know anyone who's worked in the environmental area who hasn't got a wealth of funny and salutary tales to share. The community are our eyes and ears, and once they know more about what environmental agencies do, they are very happy to call pollution hotlines and use other ways of bringing issues to our attention.

What all this really does is build the capacity of the wider community and specific stakeholder groups within it as the 'holders of the collective vision for [environmental] management. Local people are another repository of memory and vision to partner both public servants and private sector – and, where necessary, hold them accountable.'[8]

Holding the big picture and the long view in mind is increasingly important for governments and businesses all around the world, as diverse constituencies hold them to account for a sustainable future. The same applies to your environmental training programme.

Holding the big picture in mind means keeping all the elements of the Success Framework in play in proportion to each other. I've heard industry players say, for example, that, over time, the regulatory body started to focus more on compliance without giving the sector the associated support for other elements of the programme, such as industry training and engagement. Yes, you need both reactive and proactive activities – but keep them in a dynamic balance as circumstances require.

It also means holding in mind the location-specific bigger picture. To ensure that environmental training really is cost-effective – that it allocates resources in proportion to the overall importance of the issues – a holistic framework is needed. This framework needs to be developed within a wider context of integrated airshed, ecosystem, watershed and coastal management that considers both urban and rural sustainability.

Holding the long view in mind has usually been seen as the responsibility of the bodies responsible for giving effect to environmental legislation. However, over many years, I've observed these bodies being consistently held to account for how well they do this by the Māori people of New Zealand, whose world view truly takes a long-term perspective on the natural environment. Other indigenous peoples play similar roles in other countries. Increasingly, local communities, young people, environmental entrepreneurs and businesses are also stepping up to play this part. As environmental trainers, we can offer practical ways of helping governments and many other constituencies to respond to them.

 ACTION PLANNER

There is one final worksheet in the free Action Planner that accompanies this book. Use Action Sheet 7.4 to set some objectives about what to do first and to list some tasks that will help you to achieve them. Find a partner or mentor who will contact you on agreed dates to keep you on track. Also consider the longer-term future of your training programme – including, perhaps, that its success may mean you hand it on to other agencies as it becomes mainstream.

NOTES

1 Widely assumed to be a Chinese proverb – see https://proverbicals.com/roads – but also attributed to Lily Tomlin: www.brainyquote.com/quotes/lily_tomlin_379145 [both accessed 15 August 2019].

2 The law is outlined in Douglas Hofstadter (1979) *Gödel, Escher, Bach: an eternal golden braid*. Basic Books. See https://en.wikipedia.org/wiki/Douglas_Hofstadter [accessed 15 August 2019].

3 Rutherford is likely to have expressed this sentiment, although possibly in different words: see https://en.wikiquote.org/wiki/Ernest_Rutherford [accessed 16 August 2019].

4 Sourced from www.azquotes.com/author/10537-Matt_Mullenweg [accessed 15 August 2019].

5 Sourced from www.lifeethic.com/a-good-system-shortens-the-road-to-the-goal-orison-swett-marden-1850-192/ [accessed 15 August 2019].

6 In the form 'The major problem with communication is the illusion that it has occurred', this quote is widely attributed to George Bernard Shaw, but in the form cited, it appears to be from William H. Whyte (1950) Is anybody listening?, *Fortune*. See https://quoteinvestigator.com/category/george-bernard-shaw/page/4/ [accessed 15 August 2019].

7 Dr Peter M. Senge (2006) *The fifth discipline: the art & practice of the learning organization*. Currency, Doubleday, p. xvi.

8 C. Feeney and P. Gustafson (2010) *Integrating catchment and coastal management: a survey of local and international best practice*. Prepared by Parsons Brinckerhoff and Environment & Business Group for Auckland Regional Council. Auckland Regional Council Technical Report 2009/092.

Closing remarks

People who rest on their laurels are wearing them on the wrong end.
Malcolm Kushner, philosopher[1]

No successful environmental training programme can stand still. By getting on top of the easier environmental issues with our successful training, we grow our capability to grapple with the bigger existential environmental issues we face right now. This is our next mission.

This book is an instruction manual for just one tool in the toolbox for life on Earth: environmental and sustainability training. The next tray up in the toolbox is governance.

At this stage, we have an excellent array of scientific, economic and social research that has defined the scale of the issues we face and developed models to help us to visualize them. We can now conceptualize and measure our local, regional and global progress across the full spectrum of indicators of human and planetary well-being. But there is a lot of work to do to lift our partnerships to the next level that will enable local, national, international and global initiatives to dig deep and make a real difference in the next two or three decades.

The World Economic Forum (WEF), in its annual risk assessments, consistently identifies failure to mitigate and adapt to climate change and biodiversity loss and ecosystem collapse within the most damaging risk categories and the categories of risks most likely to occur.[2] Both of these risks represent major failures of governance.

The Forum's 2019 survey and report highlight concerns about 'the world's ability to deal with a growing range of collective challenges [including] the mounting evidence of environmental degradation'. The survey asked nearly 1,000 decision-makers from the public sector, private sector, academia and civil society to assess the risks facing the world and found that, 'over a ten-year horizon, extreme weather and climate change policy failures are seen as the gravest threats'. And, for the first time, after examining the human causes and effects of global risks, the report calls for greater action around rising levels of psychological strain across the world.[3]

Economists and insurance firms have, for many years, calculated that the financial costs of failure to adequately address climate change and other social and environmental issues massively outweigh the costs of taking action.[4] To me, it seems that the emotional costs of inaction are likewise starting to outweigh the costs of action.

The silly thing is that we already 'know how to solve global warming while growing our economy'.[5] Reports, such as that of the Intergovernmental Science-Policy Platform on Biodiversity and Ecosystem Services (IPBES), are releasing scientific information accompanied by summaries for policy-makers that set out detailed actions.[6] Sir Robert Watson, chair of the IPBES, says:

> It is not too late to make a difference, but only if we start now at every level from local to global. Through 'transformative change', nature can still be conserved, restored and used sustainably – this is also key to meeting most other global goals. By transformative change, we mean a fundamental, system-wide reorganization across technological, economic and social factors, including paradigms, goals and values.[7]

The moral of the story? Just do it – take action – at whatever level we operate. As our sense of agency grows, so will our capability.

I firmly believe that environmental training can light the touchpaper of individual and organizational learning, generating the transformational change that we need to transition to a happier and more sustainable world.

I wish you every success with your work.

NOTES

1 Sourced from https://izquotes.com/quote/malcolm-kushner/a-sense-of-humor-means-looking-at-things-from-an-offbeat-angle-330139 [accessed 15 August 2019].

2 World Economic Forum (2019) *The global risks report 2019*. Available at www.weforum.org/reports/the-global-risks-report-2019; see also www.weforum.org/reports/the-global-risks-report-2018 and www.weforum.org/reports/the-global-risks-report-2017 [all accessed 3 June 2019].

3 World Economic Forum (2019) *The global risks report 2019*. Available at www.weforum.org/reports/the-global-risks-report-2019 [accessed 3 June 2019].

4 (1) McKinsey & Co. (2009) *Pathways to a low-carbon economy: Version 2 of the global greenhouse gas abatement cost curve (2007)*. Available at www.mckinsey.com/business-functions/sustainability/our-insights/pathways-to-a-low-carbon-economy [accessed 3 June 2019]. (2) Lord Nicholas Stern (2007) *The economics of climate change: The Stern Review*. Cambridge University Press. (3) Oliver Ralph (2018) Insurers act on climate change exposure. *Financial Times*, 9 October. Available at www.ft.com/content/92e19630-aba2-11e8-8253-48106866cd8a [accessed 3 June 2019].

5 Natural Resources Defense Council (n.d.) A new global warming solutions map from McKinsey and Co. shows that we know how to solve global warming while

growing our economy. Media release. Available at www.nrdc.org/sites/default/
files/glo_07113001a.pdf [accessed 3 June 2019].

6 Intergovernmental Science-Policy Platform on Biodiversity and Ecosystem Services
(IPBES) (2019) *Summary for policymakers (SPM) of the global assessment report.* Find
out more at www.ipbes.net/news/Media-Release-Global-Assessment [accessed 3
June 2019].

7 Ibid.

Further resources

The following is a sample of the sources that I've found valuable for expanding my horizons about training and how we all learn. I hope you too find them helpful.

YOUR ONGOING PROFESSIONAL LEARNING
AND DEVELOPMENT AS A TRAINER

I've been a member of the New Zealand Association of Training and Development (NZATD) for about 25 years now and I learn great new stuff all the time from some of the best trainers in the world. It's part of a global network and I recommend you find a branch near you as soon as you can: it will open up amazing vistas of expertise and save you from many a mistake. Of course, our mistakes keep us learning – and so will your local association of training and development.

Associations of training and development go by many different names, and most countries will have a national association with local branches – and, hopefully, you'll find that there is one near you. Do join it.

Such associations include:

- Association for Skills Development in South Africa, www.asdsa.org.za

- Australian Institute of Training and Development, www.aitd.com.au/

- Association for Talent Development (ATD), US (formerly the American Society for Training & Development, or ASTD), www.td.org/

- Brazilian Association for Training and Development (ABTD), https://brazilsummit.td.org/

- Chinese Society for Training and Development, www.cstd.org/

- European Association of Development, Research and Training Institutes, www.eadi.org/

- Indian Society for Training & Development (ISTD), www.istd.co.in/

- Institute of Training and Occupational Learning (ITOL), UK, www.itol. org/

- New Zealand Association for Training and Development, www.nzatd.org. nz/

- Singapore Training and Development Association, www.facebook.com/ STADA.SG/

No association near you? See if there is an association of human resources (HR)/ personnel professionals or a professional speakers' association. Both groups are active in workforce training and could offer a good alternative.

FOSTERING SUSTAINABLE CHANGES IN PRACTICE

If you get interested in change, which is what training is really all about, then you may find some of the following resources helpful and inspiring:

Dr Carol S. Dweck (2017) *Mindset: Changing the way you think to fulfil your potential.* Robinson.

Dr Dweck is Lewis and Virginia Eaton Professor of Psychology at Stanford University.

Dr Doug McKenzie-Mohr (2011) *Fostering sustainable behavior: An introduction to community-based social marketing,* 3rd edn. New Society.

Les Robinson (2013) *Changeology: How to enable individuals, groups, and communities to do things they've never done before.* Scribe.

For more information, see www.enablingchange.com.au/index.php [accessed 3 June 2019].

Dr Niki Harré (2018) *Psychology for a better world: Strategies to inspire sustainability.* Auckland University Press.

Dr Niki Harré (2018) *The infinite game: How to live well together.* Auckland University Press.

Richard H. Thaler and Cass R. Sunstein (2009) *Nudge: Improving decisions about health, wealth, and happiness,* rev'd edn. Penguin.

Robert Cialdini PhD (2006) *Influence: The psychology of persuasion,* rev'd edn. Harper Business.

See a summary on Dr Will Allen's website at http://learningforsustainability.net/ post/behavior-change/ [accessed 3 June 2019].

INSPIRATION FOR CHANGING TIMES

The following are books that relate to Chapter 1 from three of my favourite economists. Highly readable, these will help you to define the scope of your environmental training mission:

Jane Gleeson-White (2015) *Six capitals – or can accountants save the planet? Rethinking capitalism for the twenty-first century.* W.W. Norton.

Find out more at https://janegleesonwhite.com/six [accessed 25 October 2019].

Tim Jackson (2017) *Prosperity without growth: Foundations for the economy of tomorrow,* 2nd edn. Routledge.

Find out more at https://timjackson.org.uk/ecological-economics/pwg/ [accessed 3 June 2019].

Kate Raworth (2018) *Doughnut economics: Seven ways to think like a 21st-century economist.* Cornerstone.

Find out more at www.kateraworth.com/ [accessed 3 June 2019].

Find out more about green jobs in these references:

Logan Yonavjak (2014) Now THIS is what we call green jobs: the restoration industry 'restores' the environment and the economy. *Forbes,* 8 January. Available at www.forbes.com/sites/ashoka/2014/01/08/now-this-is-what-we-call-green-jobs-the-restoration-industry-restores-the-environment-and-the-economy/ [accessed 3 June 2019].

K. BenDor, T. William Lester, Avery Livengood, Adam Davis and Logan Yonavjak (2013) *Exploring and understanding the restoration economy.* Available at https://curs.unc.edu/files/2014/01/RestorationEconomy.pdf [accessed 3 June 2019].

BOOKS TO HELP GET US OUT OF THE FIX WE'RE IN

For further reading relating to Chapter 2, I point you towards two books in particular:

Jared Diamond (2011) *Collapse: How societies choose to fail or succeed,* rev'd edn. Penguin.

Jared Diamond (2019) *Upheaval: Turning points for nations in crisis.* Little, Brown & Co.

A 12-step programme for nations in crisis, Professor Diamond's latest book examines how populations deal with threats to their well-being such as climate change and resource scarcity. He helps them to change for the better by means of the coping mechanisms we normally associate with personal trauma. It's just what we need! I have several of his books and, as well as being vitally informative, they are very easy to read.

You can also access an interview with Professor Diamond broadcast on Radio New Zealand on 2 June 2019 at www.rnz.co.nz/national/programmes/sunday/audio/2018697832/professor-jared-diamond-the-world-is-in-more-trouble-than-it-has-ever-been [accessed 2 June 2019].

If you want to know more about the evolution of Auckland's erosion and sediment control training programme, go to www.ESST.institute/Success/ebooks and download the free ebook, *Startled by success: The evolution of an environmental training programme.*

If you want to know more detail about each case study, go to www.ESST.institute/Success/ebooks and download the free ebook, *Same but different: Case studies of environmental training programmes.*

ELEMENTS OF SUCCESSFUL ENVIRONMENTAL TRAINING PROGRAMMES

In relation to Chapter 3, the following report is an absolute must-read for anyone working in policy and regulation – despite its dry title:

New Zealand Productivity Commission (2014) *Regulatory institutions and practices: final report.* Available at www.productivity.govt.nz/inquiries/towards-better-local-regulation/ [accessed 25 October 2019].

The following article describes how the Canadian Chapter of the International Erosion Control Association (IECA) asked the Canadian Standards Association (CSA) to produce a national standard to provide greater consistency and effectiveness in the inspection and monitoring of erosion and sediment control on construction sites. The article and standard would be informative for many other sectors:

D. O'Reilly (2019) National site erosion and sediment control monitoring standard: a groundbreaker. *Daily Commercial News*, 8 May. Available at https://canada.constructconnect.com/dcn/news/associations/2019/05/national-site-erosion-sediment-control-monitoring-standard-groundbreaker [accessed 9 May 2019].

Here are two references that list the internal and external benefits of third-party verification of environmental claims. They will also help you to identify search terms to look for such verifications for your sector:

CDP Disclosure Insight Action (2018) *The business benefits of third-party verification of climate data: A CDP guide.* Written in partnership with Bureau Veritas and Lloyd's Register. Available at http://b8f65cb373b1b7b15feb-c70d8ead6ced550b4d987d7c03fcdd1d.r81.cf3.rackcdn.com/cms/reports/documents/000/003/117/original/The-business-benefits-of-third-party-verification-of-climate_data.pdf?1520935533 [accessed 29 May 2019].

Global Reporting Initiative (GRI) (2013) *The external assurance of sustainability reporting*. Available at www.globalreporting.org/resourcelibrary/GRI-Assurance.pdf [accessed 29 May 2019].

Many environmental training programmes start out with the aim of lightening the compliance load by enhancing the capability, motivation or incentivization of firms to self-regulate. While this may not always fully eventuate, the following is a good reference to consider for your sector:

Al Iannuzzi Jr (2016) *Industry Self-Regulation and Voluntary Environmental Compliance*. CRC Press.

Al Iannuzzi is a senior environmental manager at a Fortune 500 company and examines environmental regulation through a review of compliance and enforcement theory. The book includes case studies, industry best practices, key elements and benefits.

The following book is a total must for the people responsible for implementing policies, laws and regulations. I've forgotten the name of the compliance officer who kindly recommended it to me, but in turn I strongly commend it to you:

Professor M. Sparrow (2000) *The regulatory craft: Controlling risks, solving problems and managing compliance*. Council for Excellence in Government. Brookings Institution Press.

If you want to know more about logic models, the following are some great resources:

United Nations Environment Programme/Global Programme of Action for the Protection of the Marine Environment from Land-based Activities (2006) *Ecosystem-based management: Markers for assessing progress*. UNEP/GPA. Available at http://wedocs.unep.org/handle/20.500.11822/12470 [accessed 3 June 2019].

This is Professor Stephen Olsen's order of outcomes framework adopted by the United Nations. It's elegant and comprehensive, and is designed specifically for environmental managers.

The wonderful diagram available at https://fyi.extension.wisc.edu/ programdevelopment/files/2016/03/WaterQualityProgram.pdf and the University of Wisconsin Extension report available at https://fyi.extension.wisc.edu/ programdevelopment/files/2016/03/lmcourseall.pdf [both accessed 3 June 2019].

W.K. Kellogg Foundation (2006) W.K. Kellogg Foundation logic model development guide. Available at www.wkkf.org/resource-directory/ resource/2006/02/wk-kellogg-foundation-logic-model-development-guide [accessed 3 June 2019].

P.F. McCawley (1995) The logic model for program planning and evaluation. Available at www.d.umn.edu/~kgilbert/educ5165-731/Readings/The%20 Logic%20Model.pdf [accessed 3 June 2019].

And do visit the following invaluable pages on Will Allen's website:

http://learningforsustainability.net/theory-of-change/ [accessed 3 June 2019].

http://learningforsustainability.net/logic-models/ [accessed 3 June 2019].

To find out more about participatory evaluation, the following is an excellent step-by-step guide for environmental programmes:

Jim Woodhill and Lisa Robins (1998) *Participatory evaluation for landcare and catchment groups: A guide for facilitators*. Greening Australia, Yarralumla, ACT. Available at www.academia.edu/27708397/Participatory_evaluation_for_ landcare_and_catchment_management_groups_A_guide_for_facilitators [accessed 31 May 2019].

The governance document for the Chesapeake Bay Program describes the organizational function and governance for the partnership in advancing Bay protection and restoration through the *Chesapeake Bay Watershed Agreement* signed in 2014. It is available at www.chesapeakebay.net/what/publications/23144 [accessed 3 June 2019].

You can also find out more about the huge amount of activity going on around the Bay and the strong partnerships that enable the impressive outcomes they are all achieving at the Chesapeake Bay Program partnership's website at www.chesapeakebay.net [accessed 31 May 2019].

THINGS TO PUT IN YOUR BUSINESS CASE

To help you with Chapter 4, the Harvard Business Press has published several titles on how to build a business case for any initiative, including a pocket mentor guide – for example:

Harvard Business Review (2010) *Developing a business case*. Harvard Business Press.

R. Sheen and A. Gallo (2015) *HBR guide to building your business case*. Harvard Business Press.

You might also read this very helpful book:

Dr Patricia Pulliam Phillips and Dr Jack J. Phillips (2017) *The Business Case for Learning*. HRDQ and ATD Press.

The US National Environmental Education Foundation has prepared a business case that focuses on the savings and profits that can accrue to businesses that initiate sustainability programmes. Such reports are numerous, but this one also contains several useful tips:

> National Environmental Education Foundation (2010) The business case for environmental and sustainability employee education. Available at www.neefusa. org/resource/business-case-environmental-sustainability-employee-education [accessed 5 June 2019].

Sustainability guru Bob Willard has published many great books for sustainability practitioners and has many free tools and spreadsheets on his website, https://sustainabilityadvantage.com/, including a free, open-source fill-in-the-blanks Excel workbook that monetizes all direct and indirect benefits arising from sustainability-related projects and automatically calculates ROI. See, for example:

> Bob Willard (2019) *Sustainability ROI workbook: Building compelling cases for sustainability initiatives.* Available at https://sustainabilityadvantage.com/books-dvds/roi-workbook/ [accessed 20 April 2019].

Together, the following two articles will help you to put the shine on your business case:

> Bob Willard (2016) *Three WHYs for any value proposition.* Available at http://sustainabilityadvantage.com/2016/09/02/ultbook-answering-the-big-3-whys/ [accessed 16 May 2019].

> Les Robinson (2017) *How to pitch an innovation to a risk-averse manager.* Available at https://changeologyblog.wordpress.com/2017/02/14/grumpy-manager-innovation-pitch-updated/ [accessed 30 May 2019].

I strongly recommend that you subscribe to Les and Bob's newsletters: they are not very frequent, but are always immensely valuable.

To find out more about the six capitals, the United Nations Sustainable Development Goals (SDGs), and how they can help governments and businesses, go to https://pureadvantage.org and search 'Clare Feeney' to find several relevant articles.

See also the following article by Lise Kingo for GreenBiz:

> Lise Kingo (2019) Sustainable business is good business: creating financial tipping points for the Global Goals. *GreenBiz*, 2 April. Available at www.greenbiz.com/article/sustainable-business-good-business-creating-financial-tipping-points-global-goals [accessed 7 April 2019].

Lise is the chief executive officer (CEO) and executive director of the United Nations Global Compact, a voluntary initiative based on CEO commitments to implement universal sustainability principles and to take steps to support UN goals. Find out more at www.unglobalcompact.org [accessed 7 April 2019].

I particularly commend to you this short, simple and compelling guide to integrated reporting across the six capitals for governments and businesses and the associated GRI indicator and monitoring framework:

> International Integrated Reporting Council (IIRC) (2013) *The international <IR> framework.* Available at http://integratedreporting.org/resource/international-ir-framework/ [accessed 3 June 2019].

The GRI provides a comprehensive set of indicators for monitoring outcomes across the six capitals. You can download the reporting standards for free from www.globalreporting.org/standards [accessed 3 June 2019].

Together, these two methods have been endorsed by a growing number of stock exchanges around the world that are requiring their listed companies to provide integrated reports that address ecologically and socially, as well as fiscally, responsible management.

Further helpful resources on the capital gains that can be made from restoring natural environments are:

> Whaingaroa Harbour Care, at www.harbourcare.co.nz/ [accessed 3 June 2019].

> Morrison Low (2010) *Value case for Project Twin Streams.* A report prepared for Waitakere City Council. Ref: 176103. No longer available online, but see www.morrisonlow.com/ [accessed 3 June 2019].

> D. Buchan (2007) *Not just trees in the ground: The social and economic benefits of community-led conservation projects.* WWF–New Zealand, Wellington. Available at www.harbourcare.co.nz/wp-content/files/wwfnz_not_just_trees_in_the_ground.pdf [accessed 3 June 2019].

If you are interested in environment and sustainability micro-credentials, the SustainOnline website at www.sustainonline.com/platform/ is an online academy that brings together almost 100 micro-courses through which people can learn about the fundamentals of systems-thinking and sustainability. The training platform aims to make knowledge and learning more accessible throughout entire organizations and is sponsored by The Natural Step, https://thenaturalstep.org/ [both accessed 3 June 2019].

PULLING IT ALL TOGETHER WITH ADDIE'S HELP

If you are working without the help of a professional trainer, or want to better understand what they do, then I strongly recommend this detailed step-by-step guide to help you when you reach Chapter 5:

> Michael Allen, with Richard Sites (2012) *Leaving ADDIE for SAM: An agile model for developing the best learning experiences.* American Society for Training and Development.

Other good references are:

Ron Zemke and Thomas Kramlinger (1984) *Figuring things out: A trainer's guide to needs and task analysis*. Basic Books.

Dana Gaines Robinson and James C. Robinson (2008) *Performance consulting: A practical guide for HR and learning professionals*, 2nd edn. Berrett-Koehler.

Christopher Pappas (2015) Writing learning objectives for e-learning: what e-learning professionals should know. Available at www.elearningindustry.com/writing-learning-objectives-for-elearning-what-elearning-professionals-should-know [accessed 20 May 2019].

Robin Petterd's research and application of the 70:20:10 model is truly excellent, while the Deakin White Paper gives a great overview. I warmly recommend that you visit the following pages to find out more – and even if you are not in a position to apply everything you learn, you will get excellent insight into how very good truly good training can be:

Dr Robin Petterd (n.d.) *Planting the seeds for the 70:20:10 learning model*. SproutLabs. Available at www.sproutlabs.com.au/elearning/planting-the-seeds-for-a-702010-learning-model-ebook/ [accessed 20 May 2019].

For 70:20:10 planning, tracking and evaluation, see also Robin's 'learning while working page', available at www.sproutlabs.com.au/learning-while-working/ [accessed 20 May 2019].

K. Kajewski and V. Madsen (2013) *Demystifying 70:20:10: A White Paper*. DeakinPrime. First edition January 2012; reprinted with amendments March 2012 and June 2013. Available at www.deakinco.com/media-centre/article/demystifying-70-20-10 [accessed 20 May 2019].

E IS FOR EVALUATION

For Levels 1–4 evaluation, as explored in Chapter 6, the most recent publication (at time of writing) is:

Dr James D. Kirkpatrick and Wendy Kayser Kirkpatrick (2016) *Kirkpatrick's four levels of training evaluation*. Association for Talent Development.

James and Wendy are co-owners of Kirkpatrick Partners. Together, they carry on the work of James' late father, Dr Donald Kirkpatrick. Find out more at www.kirkpatrickpartners.com/ [accessed 3 June 2019].

For Level 5 evaluation, the most recent publication is:

Patricia Pulliam Phillips (2017) *The bottomline on ROI*, 3rd edn. HRDQ.

Find out more at https://roiinstitute.net/ [accessed 3 June 2019].

Of the other publications on the ROI Institute's website, the following are great resources, including advice on how to report results and keep the sustainability initiatives going – all challenges in the real world:

> Patricia Pulliam Phillips and Jack J. Phillips (2017) *The business case for learning*. ROI Institute.

> Patricia Pulliam Phillips and Jack J. Phillips (2011) *The green scorecard: Measuring the return on investment in sustainability initiatives*. ROI Institute.

MANAGING YOUR PROGRAMME FOR SUCCESS

The following seminal references will help you to grow the learning organizations that we'll need to save the world – as you'll see in Chapter 7:

> Michael E. Porter and C. van der Linde (1995) Green and competitive: ending the stalemate. *Harvard Business Review*, Sept–Oct. Available at https://hbr.org/1995/09/green-and-competitive-ending-the-stalemate [accessed 29 April 2019].

That article was a follow-up to this one:

> Michael E. Porter (1990) The competitive advantage of nations. *Harvard Business Review*, Mar–Apr. Available at www.economie.ens.fr/IMG/pdf/porter_1990_-_the_competitive_advantage_of_nations.pdf [accessed 16 August 2019].

In 2012, the 1990 piece was found to be, by an order of magnitude, the most cited article ever in the magazine's history up to that point:

> *Harvard Business Review* (2012) Decades of influence. November. Available at https://hbr.org/2012/11/decades-of-influence [accessed 16 August 2019].

Finally, you should also seek out:

> Dr Peter M. Senge, Art Kleiner, Charlotte Roberts, Richard E. Ross and Bryan J. Smith (1994) *The fifth discipline fieldbook: Strategies and tools for building a learning organization*. Nicholas Brealey.

> Dr Peter M. Senge (2006) *The fifth discipline: The art and practice of the learning organization*, rev'd edn. Currency, Doubleday. This is a book about how companies can rid themselves of the learning 'disabilities' that threaten their productivity and success by adopting the strategies of learning organizations.

Index

Locators in **bold** refer to figures or tables

Lightning Source UK Ltd.
Milton Keynes UK
UKHW041118121219
355254UK00006B/911/P